A TENTH OF A SECOND

A TENTH OF A SECOND

A HISTORY

Jimena Canales

THE UNIVERSITY OF CHICAGO PRESS

CHICAGO AND LONDON

PUBLICATION OF THIS BOOK HAS BEEN AIDED BY A
GRANT FROM THE BEVINGTON FUND.

Jimena Canales is assistant professor of the history of science
at Harvard University.

The University of Chicago Press, Chicago 60637
The University of Chicago Press, Ltd., London
© 2009 by The University of Chicago
All rights reserved. Published 2009
Printed in the United States of America

18 17 16 15 14 13 12 11 10   2 3 4 5

ISBN-13: 978-0-226-09318-5 (cloth)

ISBN-10: 0-226-09318-2 (cloth)

Library of Congress Cataloging-in-Publication Data

Canales, Jimena.
   A tenth of a second : a history / Jimena Canales.
   p. cm.
   Includes bibliographical references and index.
   ISBN-13: 978-0-226-09318-5 (cloth: alk. paper)
   ISBN-10: 0-226-09318-2 (cloth: alk. paper)
   1. Time measurements. 2. Time—Philosophy. I. Title.
   QB213.C36 2009
   509—dc22

                                        2008054019

♾ The paper used in this publication meets the minimum
requirements of the American National Standard for Information Sciences—Permanence of Paper for Printed Library
Materials, ANSI Z39.48-1992.

To my mother, Rocío, and to my son, Billy.

CONTENTS

PREFACE

At first glance, it may seem that we can ignore the history of a tenth of a second. After all, most events occurring within this short period of time cannot be perceived. Most persons take more than a tenth of a second to react. But in looking more carefully at this moment, it appears strangely constitutive of modernity. The tenth of a second was repeatedly referenced in debates about the nature of time, causality, free will, and the difference between humans and nonhumans.

Understanding this short, "invisible" period of time is as important as understanding other equally small and invisible things. When in the seventeenth century Robert Hooke used a newly invented microscope to reveal the shocking wealth of the microworld, he claimed that the "shadow of things" no longer needed to be taken for their "substance." Microscopy led him away from "uncertainty," "mistakes," "dogmatizing," and forms of knowledge based largely on "discourse and disputation." This new technology appeared to him as important as a series of other revolutionary inventions. Hooke listed it among gun powder, the seaman's compass, printing, etching, and engraving, which together saved man from misguided attempts to advance on knowledge through wasteful "talking," "arguing," and "opining."

The idea of dividing the day into hours, an hour into sixty minutes, and a minute into sixty seconds dates back to four thousand years ago, to the contributions of the Egyptians and Babylonians. Clocks started acquiring minute hands around the year 1600. Clocks with a hand for seconds only appeared later, and the value of clocks that noted the tenth of a second only became recognized in the second half of the nineteenth century.

When in the nineteenth century scientists discovered that an analogous microworld was caught within short periods of time, they increased their technological armamentarium beyond microscopy and beyond telescopy. Just as Hooke compared the discovery of the microworld to other monumental inventions of his era, I began to compare the emergence of microtime with technologies that defined modernity. New visual technologies (such as photography and cinematography), modern media and transportation networks (such as telegraphy, steam, and rail), and contemporary theories of communication (based on concepts of signal, message, and transmission) all trafficked in microtime.

Microtime was even more pervasive than microspace since it was always changing, affecting the very essence of time and temporal development. Measurement-based science depended on understanding these moments and solving a host of new riddles that appeared with them.

I also started to ask how this discovery was used to circumvent the limitations of ordinary language already pointed out by Hooke. How did scientists fulfill the same desire that, centuries earlier, had motivated Hooke to look for alternatives to knowledge based on mere "discourse and disputation"? How were they able to install measurement as a form of knowledge distinct and superior to others? Astronomers, physicists, engineers, physiologists, and psychologists competed for answers.

In the nineteenth century, philosophers devised new theories to understand microtime, replacing the "mechanical philosophy" that characterized the Scientific Revolution. In its place they installed new ones—philosophies that would dominate the following centuries.

How can we know what happens within a tenth of a second? How do these short intervals affect history? Are there particular instances of this order of magnitude which have altered our historical landscape? What happens when we look at our history at this temporal scale? In thinking about these questions I became convinced that the challenges posed by tenth-of-a-second moments characterized modern culture from the late nineteenth to the early twentieth century. It became clear to me that central currents in modern philosophical thought could only be understood in their context. In order to rethink a longer period of time—the period known as modernity—we first need to examine what happens within these short, fugitive moments.

ACKNOWLEDGMENTS

The idea for this book started ten years ago, when I first had the opportunity to study the history of science and related fields. I was fortunate to learn about Michel Foucault from Arnold Davidson, French history from Patrice Higonnet, technology and sociology from Donald Mackenzie, European history from Charles Maier, philosophy of science from Hilary Putnam, and photography and art history from Joel Snyder. Notes that I silently wrote down during their lectures and seminars finally found a place in the pages of this book.

I want to thank my colleagues at the history of science department at Harvard University, many of whom took time from their busy schedules to read this manuscript. I am grateful that Allan Brandt, Peter Buck, Janet Browne, Anne Harrington, Everett Mendelsohn, Katharine Park, Charles Rosenberg, Steven Shapin, and Evelynn Hammonds have been sources of inspiration, support, and, of course, critique. The suggestions of Sam Schweber and Gerald Holton were invaluable for the chapters on the physical sciences.

This book could not have been finished without the constant care that Mario Biagioli and Peter Galison gave to both me and this project. My colleagues and teachers outside of the department have been equally generous, and I want in particular to thank Giuliana Bruno, Owen Gingerich, Patrice

Higonnet, Sheila Jasanoff, Charles Maier, Antoine Picon, David Rodowick, and Diana Sorensen. The attention and support of Ken Alder, Lorraine Daston, Michael Hagner, Bruno Latour, Simon Schaffer, and Norton Wise have been similarly invaluable.

I would like to thank my editor, Karen Merikangas Darling, for her advice and for flawlessly managing a thoughtful review process. My friends and colleagues helped me give meaning to every day that I worked on this project and to value every day that I did not: Peder and Nina Anker, Ladina Bezzola, Robert Brain, Marta Braun, Hasok Chang, Wendy Chun, Rand Evans, Vanesa Fernandez, Jean-François Gauvin, Mahalia Gayle, Peter Geimer, Lisa Gitelman, Jeremy Greene, Orit Halpern, Kristen Haring, Sharon Harper, Christoph Hoffman, Caroline Jones, David Kaiser, Natasha Lee, Rebecca Lemov, Karen and Brian Mariscal, Andreas Mayer, Pablo Savid-Buteler, Henning Schmidgen, Hanna Shell, Otto Sibum, Oliver Simons, Shiela and Nicholas Smithie, Laura Otis, Nuria Valverde, Heidi Voskuhl, and Andrew Warwick deserve special thanks. I also want to thank the many students who have taught me more than I have taught them, especially Bill Rankin, Jenn Clark, Alex Csiszar, Daniela Helbig, Amber Musser, Chris Phillips, Chitra Ramalingam, Miranda Mollendorf, Stephanie Tuerk, David Unger, Lambert Williams, Alex Wellerstein, and Nasser Zakariya.

Parts of this book were written while I was at Harvard University, the Max-Planck-Institut für Wissenschaftsgeschichte in Berlin, Imperial College in London, and MIT. Funding for publication was provided by the Bevington Fund of the University of Chicago Press and the Faculty of Arts and Sciences of Harvard University.

Above all, I want to thank my sisters (Lorea and Fernanda), my parents (Rocío and Ernesto), and my son (Billy), who have taught me the value of each and every tenth of a second.

CHAPTER 1

INTRODUCTION

Dwell, for a moment, on a tenth of a second. What do we know about this short period of time? What do we know about its history? Since yesterday, hundreds of thousands of tenths of a second have gone by. A year comprises hundreds of millions of these moments. How can we possibly begin to tell their history? Luckily, traces of it already exist. Many areas of science and culture have been concerned with the tenth of a second, and these accounts offer us a glimpse, albeit an oblique one, into its history.

References to the tenth of a second appeared in two main areas. The first area was predominantly scientific and technological. Starting in the mid-nineteenth century scientists associated this value with the speed of thought. This relation led them to perform reaction time experiments that increasingly defined the field of psychology.[1] Analyzing moments of the order of the tenth of a second was also central to the study of measurement errors and to the development of numerous instruments devised to

---

1. The relation between the tenth of a second, reaction time and the speed of thought, although controversial, is still widely held: "Reaction times on elementary cognitive tasks that require no conscious thought, such as responding to a lighted button, show a significant correlation with IQ scores." Charles Murray, "Afterword," in *The Bell Curve: Intelligence and Class Structure in American Life* (New York: Simon & Schuster, 1996), 561.

eliminate them. Entire disciplines, ranging from astronomy to physics, were profoundly affected by errors of this magnitude. Some of the most important measurement practices and techniques in science—from the straight lines used to divide rulers to cinematographic machines—were designed to cope with them. The second area was more broadly philosophical: references to this short period of time were part of a general inquiry into time, temporal development, and sensory experience. By extending these investigations, we can begin to use the tenth of a second as a way to rethink much longer historical periods.

The first concern was practical, and to study its history we must delve into secluded laboratory spaces, technical scientific texts, and into the minute details of one of the most mundane of scientific practices: measuring. The second concern with the tenth of a second was speculative, reflective, and open-ended. Its central purpose was to think about these moments in relation to science, technology, and modern culture. The first was narrow, the second broad. The first (scientific and technological) can be dated with precision: a concern with the tenth of a second in science intensified during the mid-nineteenth century and waned significantly by the beginning decades of the twentieth. The second one (philosophical), less so, since it continues to appear today. Both had repercussions for general culture, affecting areas beyond science and philosophy and encompassing the arts, industry, and media.

By studying the tenth of a second through these two complementary perspectives, this book reveals some of the key characteristics of the modern era and illuminates the work of some of the most important scientists and philosophers of modern times. It also offers a way to productively extend their work in new directions.

The Specificity of This Moment

The tenth of a second is, of course, not the only time interval of interest to scientists, and certainly not the shortest. In fact, since the early nineteenth century scientists had been able to measure, depict, and (later) photograph much shorter periods of time. Lensless microscopic technologies pioneered by Ernst Chladni revealed infinitely small vibrations in the complex patterns left by fine sand on vibrating plates. Similar experiments used moving mirrors to study the even smaller vibrations of tuning forks. Chronoscopes, chronographs, and myographs measured time in much smaller units, well into thousandths of a second. Photographers, first among them Henry Fox Talbot, recorded shorter periods of time by illuminating fast events with electric sparks. Others used multiple cameras, multiple lenses, or naturally intense sources of illumination. Late nineteenth- and early twentieth-century bullet, splash, and solar photography frequently went far beyond tenth-of-a-second limits. When photographic exposure time was further reduced at the end of the nineteenth century, even common subjects, including horses and birds, were captured in much smaller time periods. In 1878 Eadweard Muybridge famously photographed running horses galloping in a manner never seen before, seizing them in a 1/500th of a second by using multiple cameras. In 1882, single-lens photographs made by the French physiologist Étienne Jules Marey surpassed this limit. By 1918, Lucien Bull, the last assistant to Marey, was able to photograph fifty thousand images in one second. In the 1930s, Harold Edgerton's stroboscopic images produced at MIT captured images in the span of a 1/3,000th of a second. Today, millions of frames per second can be recorded.

In the midst of these ever-finer slices of time, the tenth of a second retained a particular salience. Tie a piece of coal to one end of a string and twirl it rapidly. If the coal makes a complete turn in less than a tenth of a second, it will seem to form a closed circle. Slow down its speed and the closed circle disappears. Try this experiment with a cinematographic projector. If successive frames pass at a speed exceeding the tenth of a second, the illusion of movement appears smoothly. Reduce its speed, and the illusion disappears. Look closely at a rapidly moving target and try to time the precise moment when it crosses a specific point. Compare the moment with somebody else's and you will see that each of your determinations will probably differ by a few tenths of a second. Step on the brake of your car when an obstacle appears in front of you, and despite your best efforts, a lag time, close to a tenth of a second, will haunt your reactions. Try to read as many words as you can in ten seconds, and you will notice that the number is about a hundred: one word every tenth of a second. Time yourself while talking, and you will see that the time needed to pronounce each syllable will never be less than a tenth of a second. Analyze the electrical rhythm of your brain, which, when at rest, will average ten cycles per second. Study a "perceptual moment" and find that it lasts about this same amount.[2]

These mundane effects, some known since antiquity and others discovered more recently, brought with them a host of philosophical riddles, paradoxes, and questions. Were they due to limitations in the processing speed of the brain, and if so, was this speed measurable? Could it be

---

2. The psychologist J. M. Stroud estimated in his paper on "The Fine Structure of Psychological Time" (1967) that a perceptual moment or "grain" of psychological time lasted about a tenth of a second.

increased or decreased? Did they point to particular physiological qualities of the human body beyond the brain, such as the speed of nerve transmission or visual persistence? How and when did these effects stray from the tenth of a second? What evolutionary function, if any, did they serve? Were they related to fundamental aspects of the universe and the cosmos? How did they veil our access to reality? Could scientists, engineers, or inventors devise technologies to overcome or alter these effects or defects? How did they transform basic beliefs about causality, movement, and history? How were they incorporated into theories of knowledge?

Answers to these questions were frequently volunteered by scientists, some of whom were influential in crafting new technologies and improving on existing ones (such as telegraphy, photography, and cinematography) to investigate these short moments of time. Yet other answers had to do with understanding the tenth of a second in relation to the much longer period known as modernity. The development of science and technology was seen as an essential characteristic of modernity, and attempts to demystify these short periods of time were considered essential to the survival of the modern project. At stake in the exceptionalism of this instant lay the exceptionalism of modernity itself. The category of modernity itself was, in a sense, parasitic on this moment.

As scientists introduced the language of modern communication theory, employing terms such as "message" and "transmission," they increasingly referred to the tenth of a second. The mid-nineteenth-century descriptions by the influential scientist Hermann von Helmholtz are characteristic: "When the message has reached the brain, it takes about *one tenth of a second,* even under conditions of most concentrated attention, before volitional transmission of the message to the motor nerves enables the muscles to execute a specific movement."[3] "Self-consciousness," he noted, lagged "behind the present" by an amount equal to "the tenth part of a second."[4] By the 1880s it was common knowledge on both sides of the Atlantic that "the time required by an intelligent person to perceive and to will is about 1/10 of a second. . . . After allowing for the time required to traverse

---

3. Cited in Johan Jacob de Jaager, "Reaction Time and Mental Processes," in *Origins of Psychometry: Johan Jacob de Jaager, student of F. C. Donders on Reaction Time and Mental Processes* (1865), ed. Josef Brozek and Maarten S. Sibinga (Nieuwkoop: B. de Graaf, 1970), 43. Italics mine.

4. Hermann von Helmholtz, "Ueber die Methoden, kleinste Zeittheile zu messen, und ihre Anwendung für physiologische Zwecke," *Königsberger Naturwissenschaftliche Unterhaltungen* 2 (1851): 325, and "On the Methods of Measuring Very Small Portions of Time, and Their Application to Physiological Purposes," *The London, Edinburgh and Dublin Philosophical Magazine and Journal of Science* 4 (1853): 189.

all of the nerves and for the latent period of the muscles, there still remains about 1/10 second for the cerebral operations."[5]

Almost a century after Helmholtz's investigations, Thomas Edison's chief laboratory engineer was amazed at how "modern communication has greatly increased the interchange of methods and ideas between the nations through the telegraph, the telephone and radio," bringing the world closer together. Yet, even then, the project had not been entirely completed. Communications were still colored by transmission delays of this same magnitude. "We all live," he concluded, "on a tenth of a second world."[6]

Why did the tenth of a second gain such importance in modernity? A psychologist would likely respond: "because of reaction time." William James, for example, referenced this value in his classic *Principles of Psychology:* "The time usually elapsing between stimulus and movement lies between one and three tenths of a second."[7] Another psychologist could then continue to explain the value of such research, ranging from the speed of thought, to improving industrial production, to simple survival in the modern world. Hugo Münsterberg, the head of the first laboratory of experimental psychology at Harvard University, described the immediate relevance of this value: "If a playing child suddenly runs across the track of the electric railway, a difference of *a tenth of a second* in the reaction-time may decide his fate."[8] If the speed of reaction could be increased by a tenth of a second, another researcher explained, we could find a way to "preempt death itself."[9]

An astronomer might say "because of the personal equation," referring to the worrisome fact that different individuals differed in their timing of star transits.[10] Referring to two well-known astronomers, the famous historian of psychology Edwin G. Boring explained the meaning of the term: "The equation, 'A − S = 0.202 sec.,' means that on average [the astronomer]

5. M. M. Garver, "Art. XXIV.—The Periodic Character of Voluntary Nervous Action," *American Journal of Science and Arts* 20 (1880): 192.
6. A. E. Kennelly, "The Metric System of Weights and Measures," *Scientific Monthly* 23, no. 6 (1926): 551.
7. William James, *Principles of Psychology* (New York: Dover Publications, 1950), 1:88.
8. Hugo Münsterberg, *Psychology and Industrial Efficiency* (Boston: Mifflin, 1913), 65. Italics mine. During a famous legal case in 1912, Karl Marbe, a student of Wilhelm Wundt, argued in favor of convicting a machinist after proving that the time he needed to avoid a railroad accident that killed fourteen people was well within the limits of his reaction time.
9. In wondering if the "speed by which the thought of resisting danger can be translated into an act" could be increased, Léon Lalanne told a story "which has become a classic" of how the Swedish king Charles XII (1697–1718), when he was hit by a mortal blow in the trenches in front of Frederikshald, fell while moving his hand toward his sword shaft. Léon Lalanne, "Note sur la durée de la sensation tactile," *Journal de l'anatomie et de la physiologie normales et pathologiques de l'homme et des animaux* 12 (1876): 453.
10. From the perspective of the history of astronomy, the most important essay dealing with the personal equation is Simon Schaffer, "Astronomers Mark Time: Discipline and the Personal Equation," *Science in Context* 2 (1988): 118.

Argelander observed transits 0.202 sec. later than [the astronomer] Struve."[11] While the *relative* personal equation compared two observers against each other, the *absolute* personal equation compared one observer's timing of an event against the time as determined by a machine. It again took the shape of an equation, where a single number was assigned to a particular person. A personal equation is rarely exactly a tenth of a second; its value tends to oscillate between one to a few tenths of a second. It is often also sometimes much more than simply an equation, since the intriguing term is often used in literary ways.

A cinematographer, on the other hand, would explain its importance in terms of the visual threshold needed for images to fuse and appear to move. Philosophers, historians, and sociologists of science would give additional answers, some of which would be central to their explanations of how we gain knowledge of the natural world.

Here lies one of the most intriguing aspects of the tenth of a second. While it appeared as a problem in the most detailed and minutely technical accounts of precision science, it was also representative of larger questions about the role of science and technology in modern culture. It was a problem that was simultaneously "of" science and "about" science—scientific and technological as well as epistemological, philosophical, and cultural.

By investigating this moment of time this book seeks to answer three questions. First, how was modernity marked by a distinctive conception of the tenth of a second? Second, how did the tenth of a second affect knowledge practices across scientific disciplines? Third, what fundamental changes in our approach to history can help us better understand the development of modern science? Allow me to elaborate.

The tenth of a second was a distinctly modern and post-Cartesian concept. When Descartes described the reaction of a person to a stimulus in his *Treatise on Man,* reaction and stimulus occurred "at the same time."

Trust in the instantaneity of nerve transmission continued into the first half of the nineteenth century. But as the century progressed, scientists started to question whether this velocity was indeed infinite. What preoccupied thinkers after 1850 was the existence of a lag time—of the order of a tenth of a second—between stimulus and response. Scientists became increasingly uncomfortable with the mind-body dualisms of Cartesian philosophy and instead focused on interfaces (such as nerves) between these two, increasingly problematic, terms.

11. Edwin G. Boring, "The Beginning and Growth of Measurement in Psychology," in *Quantification: A History of the Meaning of Measurement in the Natural and Social Sciences,* ed. Harry Woolf (Indianapolis: Bobbs-Merrill, 1961), 118.

Figure 1.1. Figure showing how the nerves conduct stimuli to the brain. Reaction and stimulus occurred "at the same time." From René Descartes, "Treatise on Man," in *The Philosophical Writings of Descartes*, 99–108 (Cambridge: Cambridge University Press, 1985), p. 102.

This change was also distinctly post-Kantian.[12] A longstanding Enlightenment and Kantian tradition treated free action as unquantifiable.[13] But from the 1850s onward, scientists increasingly turned to tenth-of-a-second measurements to explore the mechanisms of thought and of the will. By the first decades of the twentieth century Edward Bradford Titchener, founder of the first laboratory of psychology in the United States, could claim that reaction time measurements were a "crucial experiment" demonstrating that mental phenomena were measurable.[14] A focus on reaction time would reach its apogee with behaviorism. John Watson, the movement's founder,

12. In marking this period as post-Cartesian and post-Kantian, I am not claiming that these philosophies were rejected entirely. Needless to say, various forms of neo-Cartesian and neo-Kantian philosophies have continued to dominate numerous discourses.

13. Because the scientific status of a discipline depended on the amount of mathematics that it used, Kant argued that investigations on man were forever confined to certain darkness. Immanuel Kant, *Metaphysische Anfangsgründe der Naturwissenschaft* (Leipzig: Johann Friedrich Hartknoch, 1800), viii.

14. Edward Titchener traced "Wundt's attempt, by way of a single crucial experiment, to overturn the whole Herbartian psychology" to his earliest work on the personal equation. He reprinted the dialogue between Wundt and the astronomer Argelander about Bessel, which took place before the Naturforscherversammlung in 1861 at Speyer. See *Tageblatt der 36ten Versammlung deutscher Naturforscher und Aerzte in Speyer* (1861). Edward Bradford Titchener, "Notes: Wundt's Address at Speyer, 1861," *American Journal of Psychology* 34 (1923).

explained that the purpose of psychology was "to predict, given the stimulus, what reaction will take place; or, given the reaction, state what the situation or stimulus is that has caused the reaction."[15]

Traditional Kantian divisions distinguishing quantifiable from unquantifiable phenomena proved to be, if not totally incorrect, certainly overstated. Research on reaction time was used to extend quantification to the analysis of voluntary action. Quantitative sciences were no longer limited to the realm of inaction. Post-Enlightenment scientists embarked instead on the Faustian quest of extending the reach of measurement-based science to new spaces and exposing the most remote corners of the human psyche to its searching light.[16] As a result, scientists increasingly described perceptual and communication processes in terms of stimulus, message, transmission, reception, and response—essential categories that would dominate numerous discourses well into the twentieth century.

A Proliferation of Hybrids

The process of discovering what lay within tenth-of-a-second moments was fraught with social, philosophical, and even theological controversy. The issues at stake become particularly clear in comparison to the famous seventeenth-century confrontation between Thomas Hobbes and Robert Boyle on the existence of the vacuum.[17] Readers of the debate remind us that their disagreement strengthened the modern divide between the "politics of man" and the "science of things."[18] As a consequence of their confrontation, it became commonplace to think in terms of either politics or science, but rarely in terms of both. The "two cultures" organization of experience was arguably one of the most important legacies of the Scientific Revolution.

---

15. John B. Watson, *Behaviorism* (New York: People's Institute Publishing Company, 1924).
16. Michel Foucault rejected the view that it is "through some advancement in the rationality of the exact sciences that the human sciences are gradually constituted." Michel Foucault, *Power/Knowledge: Selected Interviews and Other Writings 1972–1977*, ed. Colin Gordon, trans. Colin Gordon et al. (New York: Pantheon Books, 1977), 107. About the relation of the physical and human sciences, see his explicit reference to astronomy in *Discipline and Punish*: "Slowly, in the course of the classical age, we see the construction of those 'observatories' of human multiplicity for which the history of science has so little good to say. Side by side with the major technology of the telescope, the lens and the light beam, which were an integral part of the new physics and cosmology, there were the minor techniques of multiple and intersecting observations, of eyes that must see without being seen; using techniques of subjection and methods of exploitation, an obscure art of light and the visible was secretly preparing a new knowledge of man" (p. 171).
17. Steven Shapin and Simon Schaffer, *Leviathan and the Air-Pump* (Princeton, N.J.: Princeton University Press, 1985).
18. I am referring to the reading of *Leviathan and the Air-Pump* in Bruno Latour, *We Have Never Been Modern* (Cambridge, Mass.: Harvard University Press, 1991).

The consequences of later debates about tenth-of-a-second moments are quite different. These controversies marked a new era, commonly called the Second Industrial Revolution or the Second Enlightenment.[19] This revolution was neither one of great men nor of great machines. It was characterized by science-based industrial systems based on new connections between hardware (mainly telegraphy, steam, and rail) and wetware (mainly eyes, nerves, and brains). During this period the neat division between "science" and "politics" started to be haunted by numerous problems that did not fit into these exclusive categories. Let me give one example. The relevant problem for philosophy of science was no longer principally based on long-standing questions of objectivity (or matters of fact) and subjectivity (or matters of concern).[20] Instead of struggling with the classic question of how universal, objective knowledge was different from subjective knowledge, scientists and philosophers confronted the problem of observations that varied among and within individuals. The problem was not solely of humans and nature. It was also of *humans against humans.* It was not only about *what* was true, but about *who* was right.

Nineteenth-century disputes about the tenth of a second differed in important ways from earlier scientific controversies. New philosophies focused on emerging, hybrid problems, which they solved with concepts that went beyond traditional categories of "man," "thing," "science," "society," and "politics." They heralded a new period in which these categories were precisely the elements that were called into question. A solution to the problem of knowledge involved dissolving these essential categories.[21]

Increased attention to the tenth of a second illustrated a growing concern with the temporality of cognition. Historians and philosophers have frequently focused on how, since antiquity, the intellect was modeled as a camera or mirror. Yet from early modern times onward models of cognition slowly started to shed Cartesian, static metaphors and became instead modeled after temporal ones. By the middle of the nineteenth century the temporality of cognition was widely recognized. Astronomers and physiologists were well aware that some impressions shorter than a tenth of a

19. The term "Second Enlightenment" is used in ibid.
20. For questions of objectivity and subjectivity in the sciences see Lorraine Daston and Peter Galison, "The Image of Objectivity," *Representations* 40 (1992), and *Objectivity* (New York: Zone Books, 2007). For "matters of fact" and "matters of concern" see Bruno Latour, "Why Has Critique Run Out of Steam? From Matters of Fact to Matters of Concern," *Critical Inquiry* 30, no. 2 (2004).
21. My interest in analyzing these divisions is consistent with Giorgio Agamben's claim that "it is more urgent to work on these divisions, to ask in what way—within man—has man been separated from non-man, and the animal and the human, than it is to take positions on great issues, on so-called human rights and values." Giorgio Agamben, "*Mysterium disiunctionis,*" in *The Open: Man and Animal,* ed. Werner Hamacher (Stanford, Calif.: Stanford University Press, 2004), 16.

second could simply not be perceived. Their invisibility was partly due to the surprisingly slow speed of sensorial nerve impulses (estimated to be between 25 and 65 meters per second), but the time needed by the mind to perceive, discern, and react to stimuli was also implicated. The newfound speed of thought, together with the speed of sensory transmission, caused alarming errors in astronomy, metrology, and physics. Scientists, philosophers, and even artists developed new theories of knowledge that took into account these temporal effects.

An important precedent to these investigations was established by John Locke, the famous founder of empiricism. Philosophers and historians, however, have often read Locke in the same way as they have read Descartes, focusing on their static, camera obscura model of vision. Throughout this book I will try to reverse this bias and focus instead on *temporal* cognitive models. If it can be said that Locke modeled cognition on vision, this was a type of observation that worked through time—more cinematographic than photographic. Locke, it is true, did invoke the analogy of the camera obscura to explain "understanding," but its essential features were not of image-making. Rather it was the *permanence and order of sequential images* that was essential for him. He preferred to compare the mind to a moving lantern. "Our ideas," Locke explained, "succeed one another in our minds at certain distances, not much unlike the images in the inside of a lantern, turned around by the heat of a candle."[22]

What became distinctive during the nineteenth century was the desire to measure the precise *pace* of the brain as magic lantern. It would be repeatedly analyzed and would become an object of intense debate, in astronomy and physics as much as in physiology and psychology—and the tenth of a second occupied an important place in these investigations.

Ever since Descartes popularized comparisons of humans to automata, philosophers repeatedly debated the extent to which the body could be modeled as an instrument. During the nineteenth century this question moved from being of philosophical interest to being pertinent in actual scientific practices. As scientists who investigated the tenth of a second increasingly treated the senses as instruments, they started to ask how bodily differences affected knowledge. Bodies were not the same, and this could have important repercussions for science. If instruments were com-

---

22. John Locke related the smallest unit of time that could be perceived to other units of perception (of length, angles, color, volume, weight, tone, etc.), noting that angular measures smaller than one-hundredth of a degree could simply not be perceived. John Locke, *An Essay Concerning Human Understanding* (London: Penguin Books, 1997), 177.

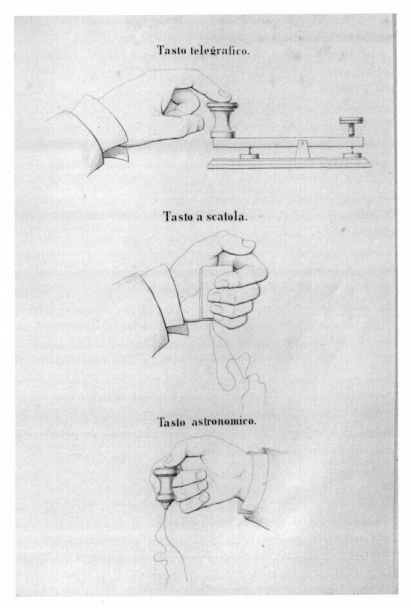

Figure 1.2. A comparison of recording keys used in different areas. From P. Tacchini, "Sulla equazione personale," *Rivista Sicula di scienze, letteratura ed arti* 2 (1869): 382–92, after p.392.

pared to the senses, how could one be assured of the precise moment when they ended and perception began? Questions of time and its relation to space became entangled with questions relating to bodies, body types, and body parts.

The difficulty in establishing a boundary between experimenters and instruments became particularly evident in the study of new scientific objects that were seen to travel through space as sensations traveled through nerves. These included gravitational and magnetic effects, light, electricity, and sound (called, in the nineteenth century, "the invisible telegraph used by nature").[23] Experiments on light were so intertwined with nerve theory that scientists found theories of light to "have their counterpart in the physiology of nerves."[24] Measurements of heat using thermometers and of electric currents with galvanometers were increasingly clouded by precisely the question of exactly where and when did nervous transmission end, and where, when, and how did perception take place? Even different inscription keys affected the result of experiments. Which of these should they adopt?

Measurement—one of the most basic operations of science—was affected by debates about the tenth of a second. Did astronomers and physicists need to study the speed of nervous transmission, nerve lengths, and the speed of thought? While tenth-of-a-second delays made patent scientists' need to take into consideration physiological velocities, and while slight differences in this number made it clear that they needed to be aware of differences among individuals, scientists increasingly despaired over where to end their investigations.

Measurement Compared to Discourse

How is the process through which scientists arrive at scientific truth distinct from (and superior to) processes through which we achieve consensus in other human endeavors? How is it different from the way others, say, politicians, reach agreement through common discursive practices? It is indeed tempting to conclude that we have arrived at our knowledge of the tenth of a second by following a special route, which has nothing to do with politics, with common discursive practices, or with other human concerns. It is equally tempting to claim that new "modern" types of knowledge are

23. Rodolphe Radau, *L'Acoustique ou les phénomènes du son*, Bibliothèque des merveilles (Paris: L. Hachette et Cie, 1867), 1.
24. Georges Pouchet, "Le Système nerveux et l'intelligence," *Revue des deux mondes* 93 (15 June 1871): 723.

stronger than their premodern and primitive alternatives. It is only natural to believe that measurement-based science is the sturdiest of these modern forms of knowledge. Yet these conclusions appear quite differently if we follow the history of the tenth of a second. The view that science was an exceptional form of knowledge became widespread during this time partly because of novel conceptions of the tenth of a second.

A careful attention to the tenth of second reveals a number of debates behind the silent facade of measurement traditionally portrayed by historians and philosophers. Studying these debates offers a way to rethink theories of modernity in which measurement remains mostly unthought of. While we are routinely fascinated by political actions, institutional histories, and the roots and consequences of irreconcilable ideologies, we have neglected how agreement works on a more basic level, for example, in measurement. How do scientists determine the length of a ruler, the moment of contact between celestial bodies, the speed of fugitive events and occurrences within a tenth of a second? How are these low-level measurement practices comparable to other ones in which individuals also strive to reach consensus? How are they used to uphold modern dichotomies, such as the spatiotemporal divide between moderns and primitives and the cognitive divide between rationality and irrationality?

A history of brief time goes against the grain of Enlightenment master narratives where quantification appears as one of the salient features of modern progress.[25] My object of study may appear much less momentous than most. This book features few great men, no lone geniuses, only small explosions, and no exotic expeditions.[26] The Franco-Prussian War and WWI appear on the periphery, tenuously connected to the problem of the tenth of a second through a concern with aiming and reacting. Revolutions, wars, coups d'état, and political struggles appear only when they affected scientists, their bodies, and their body parts.

As part of a widespread reevaluation of the Enlightenment that emerged in the late 1960s, historians and philosophers have started to question "master" quantification narratives. Instead of seeing measurement as a cumulative and progressive process, a number of scholars have noticed how the

---

25. In this respect, I agree that the question of science can no longer be considered apart from the question of modernity. See Joseph Rouse, "Philosophy of Science and the Persistent Narratives of Modernity," *Studies in History and Philosophy of Science* 22, no. 1 (1991).

26. In this respect, my object of study differs from the common focus on well-known scientific controversies that has concerned most work in the Sociology of Scientific Knowledge. Harry Collins and Trevor Pinch, for example, believe that "citizens as citizens need understand only controversial science." See Harry Collins and Trevor Pinch, *The Golem: What You Should Know about Science* (Cambridge: Cambridge University Press, 1998), xv.

accumulation of measurements is meaningless if unaided by theoretical presuppositions. More recent historians have shown how the Enlightenment
"quantifying spirit" was not limited to science, but part of a broader "avalanche of numbers" essential to accounting practices, insurance companies,
meritocratic institutions, military cultures, population surveys, taxation,
land administration, and policing.[27]

The elucidation of the social and historical nature of quantification pro-
vides a solid base for a *different* type of exploration. Instead of focusing
on local political and social aspects of modernity that affected the place of
numbers in society, this book is centered on *the moment of measurement*.
All measurements (including "measurements of distance") require a "making present" that is intimately connected to problems of a temporal order.[28]
Instead of studying the tenth of a second *in* modernity, my aim is to understand the tenth of a second *as* modernity. This period was marked by
three new challenges. How were boundaries among humans and between
humans and nature established? How was measurement produced by individuals but held free from individual biases? What were the respective
merits and the proper roles of philosophy, history, and science?

Direct References, Scope, and Organization

Most *direct references* to the tenth of a second in science appear in four
principal areas: experimental psychology (reaction time); the exact sciences
of astronomy, physics, and metrology (personal equation and measurement
errors); physiological optics (visual persistence); and imaging technologies
(graphic, photographic, and cinematographic). All of these areas have rich
historiographic traditions. Since these accounts touch, although indirectly,
on this interval, any account of the tenth of a second must intersect with

27. The "quantifying spirit" is the title and driving concept of Tore Frängsmyr, J. L. Heilbron, and Robin
R. Rider, eds., *The Quantifying Spirit in the Eighteenth Century* (Berkeley: University of California Press,
1990). It is also mentioned in William Clark, Jan Golinski, and Simon Schaffer, "Introduction," in *The
Sciences in Enlightened Europe*, ed. William Clark, Jan Golinski, and Simon Schaffer (Chicago: University
of Chicago Press, 1999). The term "avalanche of numbers" is from Ian Hacking, "Biopower and the Avalanche of Numbers," *Humanities and Society* 5 (1983).
28. It is important to highlight that my preoccupation is not with scientific determinations *of* time but
rather that my concern lies with scientific operations as they occur *in* time. "Measuring time is essentially
such that it is necessary to say 'now'; but in obtaining the measurement, we, as it were, forget what has
been measured as such, so that nothing is to be found except a number and a stretch." Martin Heidegger,
Being and Time (New York: Harper-Collins, 1962 [1927]), 471. While Heidegger did not expand on time
measurement in astronomy, he did note the need for a deeper inquiry: "The connections between historical numeration, world-time as calculated astronomically, and the temporality and historicality of Dasein
need a more extensive investigation." Ibid., 499 n. iv.

them. Studying the entanglement of these fields can help us delineate the contours of late nineteenth- and early twentieth-century science and reveal how elite modern Western culture grappled with the experience of time and its fragmentation—and our sense of history.

Every chapter of this book starts with direct references to this value. Each then branches out to elucidate much broader connections. The first chapter after the introduction focuses on reaction time and the astronomical problem of the personal equation, two areas where the tenth of a second was frequently referenced. It starts by recounting the "standard account" of the history of these topics, which is usually written from the perspective of the history of astronomy and experimental psychology.[29]

In the well-known and often-repeated "standard account" of the history of experimental psychology, the discovery of reaction time and the personal equation heralds a revolutionary era marked by the eclipse of the theatrical methods used by Jean-Martin Charcot, the famous psychologist known for his studies of hysteria, and by an increase in brass-instrument and laboratory-oriented techniques. While standard accounts have centered largely on the contributions of the aforementioned Helmholtz (as the first person to successfully measure the speed of sensory transmission in humans), the Dutch scientist Franciscus Cornelis Donders (as the first scientist to successfully measure the speed of thought), the German scientist Wilhelm Wundt (founder of the first laboratory of experimental psychology), the British astronomer royal Nevil Maskelyne (as the first person to notice the personal equation), and the German astronomer Friedrich Bessel (known, mistakenly, as the first to introduce the term "personal equation"), this chapter examines previously undisclosed fissures and debates about how this history should be told and describes alternatives. In moving beyond the traditional account of these topics, it sketches a broader context of both reaction time and the personal equation by using selective examples across a broad range of time periods and nations. It takes us through France, Germany, Britain, and North America, focusing on important figures (Karl Pearson, Frederick Winslow Taylor, Carl G. Jung, Wilhelm Ostwald, and Aldous Huxley).

Chapter 3, "The Measure of All Thoughts," examines how the tenth of a second emerged as an elemental unit of human consciousness. It claims

---

29. A typical assessment of the personal equation exclusively in terms of astronomy and experimental psychology is Kathryn M. Olesko, "Error & the Personal Equation," in *Oxford Companion to the History of Modern Science*, ed. J. L. Heilbron (Oxford: Oxford University Press, 2003), 272. She claims that other than in the astronomers' work on the personal equation, "it was seldom used elsewhere except in psychology."

that its importance derived from a new experimental system comprised of stimulus, subject, and recording device. This system would increasingly characterize many areas of modern science from psychology to cybernetics. It became a model for thinking about our engagement with the world and provided the essential categories for twentieth-century semiotic theories.

The next chapter focuses on "the astronomical event of the century": the nineteenth-century transit of Venus and the errors that plagued its observation. While observation and measurement errors were rarely acknowledged in scientific texts addressed to a broad public, they abounded in scientific accounts that were highly technical, circulated to a small group of peers, or almost entirely private. In this chapter, it becomes essential to explore these often neglected sources (hidden in archives, research and private notebooks, and scientific and journalistic articles) to reveal intrigues that were often left out of more widely circulated texts. On this exceptional occasion, debates about tenth-of-a-second measurement errors could not be kept under wraps, motivating scientists to experiment with new visual photographic and cinematographic techniques in the hope that, with them, they would eliminate these errors. Yet these debates were not limited to astronomy. Other agonistic struggles, including important military conflicts of the modern period, were frequently understood in terms of stimulus and reaction.

Chapter 5 delves deeper into the histories of photography and cinematography. Research on these technologies was often justified by how they related to tenth-of-a-second moments. The Lumière camera of 1895, by recording up to 18 frames per second, captured and projected scenes *within* a tenth of second. These cinematographic advances changed the topography of science, where it started to differ from what had characterized it for the preceding fifty years. At the very moment the cinematographic instrument on which many scientists had pinned their hopes for recording this moment was found, this machine split in two, into the cinematographic camera and the cinematographic projector.[30] This split entailed changes in how the observer was thought of by both the exact and life sciences. New conceptions of the tenth of a second emerged along with fun-

---

30. For Gilles Deleuze, an essential part of "cinema's position at the outset" was that "the apparatus for shooting [*appareil de prise de vue*] was combined with the apparatus for projection, endowed with a uniform abstract time." For his purposes this moment was important only when it was surpassed. Throughout his work he is concerned with the "emancipation of the point of view, which became separate from projection." In contrast, I focus on the previous period characterized by the combination of shooting and projection apparatus. Gilles Deleuze, *Cinema 1: The Movement-Image*, trans. Hugh Tomlinson and Barbara Habberjam (Minneapolis: University of Minnesota Press, 1986), 3.

damental transformations in the relation between art, science, and media technologies.

Chapter 6 tracks the tenth of a second as it affected the development of the exact sciences. It traces its history from nineteenth-century astronomical observatories to twentieth-century laboratories of physics and explores how it entered into debates about the establishment of the new discipline of theoretical physics. This fundamental transformation was welcomed by a new generation of scientists such as the physicists Max Planck and Albert Einstein, but it was resisted by a previous generation led by the radical physicist-philosophers Ernst Mach and Pierre Duhem. Mach and Duhem espoused a view of science in which its connections with history and philosophy were paramount and which differed markedly from later movements that stressed its autonomy. For them, the concept of the personal equation was essential to their critiques of positivism. By referring to the personal equation, Duhem built a case against the concept of "crucial experiment," and Mach stressed the conventional nature of common divisions between physiological and physical elements.

The chapter before the conclusion deals with a central debate in twentieth-century history and philosophy of science. It starts in 1922, when short moments of time were discussed during the famous confrontation between the French philosopher Henri Bergson and Einstein on the meaning of relativity. While Bergson and others tried to think of the theory of relativity in light of these moments, Einstein successfully managed to exclude them. Einstein's widely perceived "victory" against Bergson heralded a new era, where physics could advance unfettered by the questions that had so greatly preoccupied late nineteenth- and early twentieth-century philosophy. By this time, most astronomers and physicists employed a variety of new instruments designed to correct for tenth-of-a-second measurement errors, including "impersonal" micrometers and interferometric techniques, hailed in the twentieth century as the crowning achievement of measurement-based science. Yet the problem was not entirely solved. Even after the proliferation of these instruments, even after the invention of cinematography, and even after his debate with Einstein, Bergson continued to argue that real movement, real change, and real events escaped *between the static intervals of time* used in the sciences. His philosophy offers a way of thinking about scientific progress beyond its frequent portrayal in term of problems and solutions, even when technological advances and new instruments played important roles.

By the end of the 1920s debates in science pertaining to tenth-of-a-second moments had largely disappeared. Even the astronomers' problems had been mostly resolved. For researchers who worked in the early decades of the twentieth century, new instruments and techniques had completely demystified the tenth of a second. Controversies pertaining to this moment, however, did not completely vanish. The implications of tenth-of-a-second debates for science were not entirely forgotten even after numerous solutions to the problem of the personal equation were implemented in astronomy, physics, and metrology, even after the invention of new imaging techniques for visualizing short periods of time proliferated, and even after reaction time research became part of the experimental psychologist's daily work. When philosophies contesting logical positivism started to gain prominence after WWII, some of these relied on alternative interpretations of tenth-of-a-second moments.

The conclusion explores how our view of knowledge production needs to change in light of a history of the tenth of a second. It first delves into the work of the historian of science Thomas Kuhn, author of the *Structure of Scientific Revolutions,* and shows how historians, sociologists, and philosophers after Kuhn rethought the nature of measurement and quantification. It then covers two examples from the philosophy of science with direct references to the personal equation: one from the philosopher Karl Popper, author of the monumental *Logic of Scientific Discovery,* and a second one from the famous chemist and philosopher Michael Polanyi. The works of Popper and Polanyi established the reigning norms of later "antipositivist" work in the history and philosophy of science, leaving a deep mark on later historiography. Popper turned directly to the personal equation to combat some of the central assumptions of logical positivism. He reminded us of the intersubjective and social elements of scientific production. Polanyi invoked tenth-of-a-second moments to hone his controversial account of science, stressing its personal nature. Through these examples we can see how studies on the tenth of a second entered into debates about reference and signification, about how statements connect to the world. We can see how the analysis of this interval was central to the century-long attempt to establish a semiotic theory of meaning.

The conclusion branches beyond the history and philosophy of science to further explore the tenth of a second. It asks what it means to write the history of this time period, as opposed to writing about a year, a decade, or a century. It then explores how our understanding of the tenth of a second arose in the context of more general structures of agreement and consensus in modernity.

The postscript explores the famous comment by the critic Walter Benjamin on the "dynamite of the tenth of a second." I conclude with one lesson: the need to rethink this moment of time in philosophy, in history, in culture, and in science.

The governing idea behind this book is that modern conceptions of the tenth of a second are essential for understanding boundaries in modern life—reality and illusion, action and passivity, psychology and physiology, the human and the nonhuman, as well as the mechanical versus the spiritual—boundaries felt in bodies considered variously as sensors, reactors, authors, and texts.[31]

---

31. In considering bodies as sensors, reactors, authors, and texts, I am drawing on the work of the philosopher Gilles Deleuze. Deleuze turned to the tenth of a second to rethink the concept of authorship and texts in terms of association velocities. He argued that our traditional conception of reading was based on the idea that it occurred at "vitesse = o," rendering the meaning of a text stable. Deleuze experimented with altering this speed in order to question a text's stability. He admired forms of writing, particularly those pioneered by Hélène Cixous, meant to be read at "greater and greater speeds . . . according to the speed of association of the reader" reaching all the way to a "tenth of a second." One of the advantages of these literary experiments was that different meanings appeared at different speeds. They proved to him that texts did not reflect a frozen or established reality but rather one of change and possibility. In this way, static reading practices could be replaced by dynamic coauthorships between writers and readers. Gilles Deleuze's views about the effects of the tenth of a second on reading and writing illustrate how this value can still be used to alter common views on meaning-making, both textual (in terms of communication) and natural (in terms of knowledge). Gilles Deleuze, "Hélène Cixous ou l'écriture stroboscopique," in *L'Île déserte et autres textes: Textes et entretiens 1953–1974*, ed. David Lapoujade (Paris: Les Éditions de Minuit, 2002 [1972]), 321.

CHAPTER **2**

REACTION TIME AND THE
PERSONAL EQUATION

We should always remember the famous case of the Astronomer Royal,
Nicholas [*sic*] Maskelyne, who dismissed his assistant Kinnebrook for
persistently recording the passage of the stars more than half a second
later than he, his superior.

MICHAEL POLANYI, chemist and philosopher

A widely read and influential text by the famous psychologist and historian
of psychology Edwin G. Boring (1929) described a century-long obsession
with the "sacred 0.1 sec."[1] When did this value first appear as a "sacred," salient, and troubling, time period? According to Boring, the incident that
triggered this devotion was well known to "every psychologist."[2] The date
was 1796, and the place was the Royal Observatory in Greenwich. That year
David Kinnebrook, an assistant to the British and Royal Astronomer Nevil
Maskelyne, was fired because his observations systematically differed from
those of his superior.[3] An influential author writing decades after Boring was

1. Edwin G. Boring, *A History of Experimental Psychology* (New York: Appleton-Century-Crofts, 1929), 148.
2. Ibid.
3. For repetitions of the Maskelyne incident see Rodolphe Radau, "Sur les erreurs personnelles," *Le Moniteur scientifique* 7 (1865): 977; Etienne-Jules Marey, "Leçon d'ouverture: Vitesse des actes nerveux et

even stricter about its historical origins: "Any history of the personal equation in astronomy *must necessarily* begin" at that time.[4] In *Personal Knowledge* (1958), the famous chemist and philosopher Michael Polanyi commenced his account of science with this same sage tale: "We should always remember the famous case of the Astronomer Royal, Nicholas Maskelyne, who dismissed his assistant Kinnebrook for persistently recording the passage of the stars more than half a second later than he, his superior."[5]

The tenth of a second was closely associated with the discovery of the personal equation in astronomy and reaction time in experimental psychology. Its history thus forms part of a "standard account" of the history of experimental psychology that has been repeated for more than a century.[6]

The personal equation and reaction time were two controversial terms whose exact meaning would be debated for decades. The term "reaction time" was mostly used by experimental psychologists to describe a lag time, of the order of the tenth of a second, between stimulus and response; the term "personal equation" was mostly used by astronomers. Different astronomical observers assessed time differently, and while these assessments showed a remarkable constancy within the same person, when individuals were compared against each other, results often varied by a few tenths of a second. Many astronomers believed that one reason why observers differed in these estimations was due to their different times of reaction. As

cérébraux.—Le Vol dans la série animale, Collège de France, histoire naturelle des corps organisés, cours de M. Marey," *Revue scientifique* 6, no. 4 (26 December 1868): 62; Sigmund Exner, "Experimentelle Untersuchung Der Einfachsten Psychischen Processe. Erste Abhandlung: Die Persönliche Gleichung," *Archiv für die gesammte physiologie des menschen und der thiere, herausgegeben von Dr. E. F. W. Pflüger* 7 (1873): 602. Henri Beaunis, *Nouveaux éléments de physiologie humaine comprenant les principes de la physiologie comparée et de la physiologie générale* (Paris: J.-B. Baillière et fils, 1876), 1030; Etienne-Jules Marey, *La Méthode graphique dans les sciences expérimentales et principalement en physiologie et en médecine* (Paris: G. Masson, 1878), 144; Henri Beaunis, *Nouveaux éléments de physiologie humaine comprenant les principes de la physiologie comparée et de la physiologie générale*, 2nd ed. (Paris: J.-B. Baillière et Fils, 1881), 2:1363, and *Nouveaux éléments de physiologie humaine comprenant les principes de la physiologie comparée et de la physiologie générale*, 2nd ed. (Paris: J.-B. Baillière et fils, 1888), 2:801; A. Rémond, *Recherches expérimentales sur la durée des actes psychiques les plus simples et sur la vitesse des courants nerveux à l'état normal et à l'état pathologique* (Paris: Octave Doin, 1888), 11; Edmund C. Sanford, "Personal Equation," *American Journal of Psychology* 2 (1889): 8; Joseph Jastrow, *The Time-Relations of Mental Phenomena*, Fact and Theory Papers (New York: N. D. C. Hodges, 1890), 21; Paul Fraisse, "The Evolution of Experimental Psychology," in *History and Method*, ed. Jean Piaget, Paul Fraisse, and Maurice Reuchlin, vol. 1 of *Experimental Psychology: Its Scope and Method*, ed. Jean Piaget, Paul Fraisse, and Maurice Reuchlin (New York: Basic Books, 1968).

4. Raynor L. Duncombe, "Personal Equation in Astronomy," *Popular Astronomy* 53 (1945): 2–13, 63–76, 110–21. Italics mine.

5. Michael Polanyi, *Personal Knowledge* (Chicago: University of Chicago Press, 1958), 19–20.

6. Boring, *A History of Experimental Psychology*, 133. Some examples are Fraisse, "The Evolution of Experimental Psychology"; G. P. Brooks and R. C. Brooks, "The Improbable Progenitor," *Journal of the Royal Astronomical Society of Canada* 73 (1979). For a critique of the "standard account," see Christoph Hoffmann, *Unter Beobachtung: Naturforschung in der Zeit der Sinnesapparate* (Göttingen: Wallstein Verlag, 2006).

the terms reaction time and personal equation gained currency, their definitions nonetheless remained in flux well into the twentieth century.

One of the first attempts to differentiate their meaning came from the
famous Viennese scientist Sigmund Exner, who in the 1870s, coined the
phrase "reaction time" to distinguish it from "personal equation," "personal
time," "personal error," and "individual differences."[7] Yet scientists after
Exner continued to debate and argue about their differences and similarities. What is more, the use of these terms often exceeded the boundaries
of science, affecting philosophy. They even became labels for biases, prejudices, and personal subjectivities.

The unfair dismissal of Kinnebrook by Maskelyne showed how excessive power could corrupt knowledge. The incident became a rallying symbol
of a small but sturdy revolutionary movement within science whose goals
echoed those of the larger European revolutions. The story was told using
politically charged language. It was (so the story went) a moment when the
authoritarian Astronomer Royal unfairly repressed his assistant's unique
way of seeing. Tenth-of-a-second differences could only be acknowledged
in scientific spaces organized differently, where subordinate observers were
not so easily dismissed. According to some, the lesson to be drawn from
this incident was that science, like society, was also hurt by authoritarian
regimes. Excessive power marred scientific results. In France, it challenged
the ideals of *égalité, fraternité,* and *liberté* in national and local terms. Certain scientists of dangerous Napoleonic ilk, such as Urbain Le Verrier, director of the Paris Observatory during the second half of the century, and
Jean-Martin Charcot, the expert in hysteria who influenced a generation of
psychologists, threatened these values. A group of rebellious scientists rose
against their superiors, advocating new scientific practices characterized by
friendly and exchangeable roles between astronomers and observers and
subjects and experimenters. They stressed the benefits of new laboratory-
oriented practices characterizing an emergent discipline that they came to
name "experimental psychology." This new field stood in sharp contrast
to traditional clinical psychology, where dismal power differentials existed
between scientists such as Charcot, his mostly female patients, and even his
students and colleagues.

7. Sigmund Exner, "Experimentelle Untersuchung der einfachsten psychischen Processe. Erste Abhandlung: Die persönliche Gleichung," *Archiv für die gesammte Physiologie des Menschen und der Thiere, herausgegeben von Dr. E. F. W. Pflüger* 7 (1873): 608–9. See also Sigmund Exner, "Experimentelle Untersuchung
der einfachsten psychischen Processe. Zweite Abhandlung: Ueber Reflexzeit und Rückenmarksleitung,"
Archiv für die gesammte Physiologie des Menschen und der Thiere, herausgegeben von Dr. E. F. W. Pflüger 8
(1874): 526–37.

Histories of the astronomers' discovery of the personal equation conveyed various lessons. Connected to social and political ideals, scientists claimed that this moment marked the beginning of a new era, where the philosophies of Descartes and Kant would finally be surpassed, where scientific materialism would eliminate the last vestiges of religious irrationality, and where new forms of political representation reigned supreme.[8] The "standard account" was particularly important because it was a parable demonstrating a balanced (neither too excessive nor too narrow) use of measurement in modernity.

While debates about the significance and meaning of the personal equation and reaction time concerned much more than experimental psychology and astronomy, historians have nonetheless focused on them through the lenses of these two disciplines. They have also given a consistent view of their origin.

The "Standard Account" from Bessel to Wundt

According to the "standard account," the Prussian astronomer Friedrich Bessel was the first scientist after Maskelyne to study the personal equation, starting in 1815.[9] Investigations then remained dormant for a number of years, partly due to a longstanding belief in the impossibility of measuring the speed of sensory transmission. Johannes Müller, the doyen of German physiology, considered the speed of nerve transmission to be immeasurable. In the *Elements of Physiology* (1844), he explained:

> The attempts made to estimate the velocity of nervous action have not been founded on sound experimental procedures. Haller calculated that the nervous fluid moved with the velocity of 9,000 feet in a minute; Sauvages estimated the rate of its motion at 32,400, and another physiologist, at 57,600 million feet in a second. . . . We shall probably never attain the power of measuring the velocity of nervous action; for we

---

8. For recent work on reaction time experiments and French experimental psychology see Henning Schmidgen and Jacqueline Carroy, "Reaktionsversuche in Leipzig, Paris und Würzburg: Die deutsch-französische Geschichte eines psychologischen Experiments, 1890–1910," *Medizinhistorisches Journal* 39 (2004); Jacqueline Carroy, "Théodule Ribot et la naissance d'une psychologie scientifique," in *L'Anhédonie: Le Non-Plaisir et la psychopathologie*, ed. M.-L. Bourgeois (Paris: Masson, 1999); Jacqueline Carroy and Régine Plas, "The Origins of French Experimental Psychology: Experiment and Experimentalism," *History of the Human Sciences* 9, no. 1 (1996): 77.

9. The original is Wilhelm Friedrich Bessel, "Personal Gleichung," *Astronomische Beobachtungen* 8 (1822).

have not the opportunity of comparing its propagation through immense space, as we have in the case of light.[10]

This state of affairs started to change in the 1850s, when the German scientist Hermann von Helmholtz and the physiologist Emil du Bois-Reymond compared the nervous agent to electricity and assigned a finite speed to both.[11] Their canonical experiments overturned the longstanding belief in the instantaneity of nerve transmission. In the following decades, an increasing number of scientists started comparing the nervous system to emerging telegraph networks.[12]

Helmholtz and du Bois-Reymond argued that delays in nerve transmissions overwhelmed physical quantities by their magnitude. While physicists thought they were competently dealing with speeds of more than 400 million meters per second, they often ignored that their own organisms functioned at a much slower speed of 26–30 meters per second. By refusing to consider physiological elements, they argued, scientists introduced enormous errors into their results. Among their examples, they included experiments on the speed of electricity, light, sound, and the velocity of the earth, as well as mundane examples of speeding horses—of military, commercial, scientific, and artistic concern. Even astronomers, they insisted, could "profit" from the findings of physiologists.

After intially working with frogs, Helmholtz suggested, but never entirely succeeded in, a way to extend his spectacular research to humans. From his examinations of humans, he concluded that "the quickness of reflection is . . . by no means so great as seems to be assumed in the expression

---

10. Johannes Müller, *Elements of Physiology*, 2 vols., trans. William Baly (London: Taylor and Walton, 1838–42), 729. Originally published as Johannes Müller, *Handbuch der Physiologie des Menschen* (Coblentz: Hölscher, 1834), 1:581.

11. For Helmholtz's work with frogs and the graphic method, see Frederic L. Holmes and Kathryn M. Olesko, "The Images of Precision: Helmholtz and the Graphical Method in Physiology," in *The Values of Precision*, ed. M. Norton Wise (Princeton, N.J.: Princeton University Press, 1995); Kathryn M. Olesko and Frederic L. Holmes, "Experiment, Quantification, and Discovery: Helmholtz's Early Physiological Researches, 1843–50," in *Hermann von Helmholtz and the Foundations of Nineteenth-Century Science*, ed. David Cahan (Berkeley: University of California Press, 1993); Robert M. Brain, "The Graphic Method: Inscription, Visualization and Measurement in Nineteenth-Century Science and Culture" (Ph. D. thesis, University of California, Los Angeles, 1996); Robert M. Brain and M. Norton Wise, "Muscles and Engines: Indicator Diagrams and Helmholtz's Graphical Methods," in *The Science Studies Reader*, ed. Mario Biagioli (New York: Routledge Press, 1999).

12. For comparisons between telegraph networks and the nervous system, see Christoph Hoffmann, "Helmholtz' Apparatuses: Telegraphy as a Working Model of Nerve Physiology," *Philosophie Scientiae* 7 (2003); Iwan Rhys Morus, " 'The Nervous System of Britain': Space, Time and the Electric Telegraph in the Victorian Age," *British Journal for the History of Science* 33 (2000).

Considérations suggérées par les résultats précédents. — La table suivante offre l'occasion de comparer la vitesse de l'agent nerveux, telle qu'elle a été établie par les recherches précédentes, avec celle de plusieurs autres agents ou corps en mouvement, et l'on peut tirer de cette comparaison plusieurs conclusions intéressantes :

Vitesse de	Mètres en une seconde.
Électricité (expériences de M. Wheatstone)...	464 000 000
Lumière.............................	300 000 000
Son dans le fer.....................	3 485
— dans l'eau.......................	1 435
— dans l'air.......................	332
Étoiles filantes....................	64 380
Terre dans son orbite autour du soleil.......	30 800
Surface de la terre à l'équateur...........	465
Boulet de canon.....................	552 (1)
Vent..............................	1—20
Vol de l'aigle.......................	35 (2)
Locomotive.........................	27
Lévrier, cheval de course.............	25

Figure 2.1. Chart of common velocities against which du Bois-Reymond compared the speed of nerve transmission. From Emil du Bois-Reymond, "Vitesse de la transmission de la volonté et de la sensation à travers les nerfs: Conférence de M. du Bois-Reymond à l'Institution Royale de la Grande-Bretagne," *Revue scientifique* 4, no. 3 (15 December 1866): 33–41, on p. 38.

'quick as thought.'"[13] A commentator on his work remarked how "under the most favorable conditions and with a highly sustained attention, the brain needs at least 0.1 seconds for transmitting its orders to the nerves which conduct voluntary movements."[14]

In the eyes of physiologists, Helmholtz's work heralded a new physicalism that displaced a vitalism based on the concept of the *Lebenskraft* or "life force." For physicists, his research on the conservation of force was equally revolutionary. They claimed it as one of the pillars of the new science of thermodynamics. Helmholtz's results soon became world famous. A commentator explained the implications of his research. If the personal equation was related to the speed of thought, and if it could be reduced through

13. Hermann von Helmholtz, "Ueber die Methoden, kleinste Zeittheile zu messen, und ihre Anwendung für physiologische Zwecke," *Königsberger Naturwissenschaftliche Unterhaltungen* 2 (1851): 325, and "On the Methods of Measuring Very Small Portions of Time, and Their Application to Physiological Purposes," *The London, Edinburgh and Dublin Philosophical Magazine and Journal of Science* 4 (1853): 189.
14. Otto Eduard Vincenz Ule, "Sur les moyens de mesurer la pensée: Lettre de M. Ule à M. E. Desor," *Revue suisse* 20 (1857).

education, could someone become more intelligent through practice and discipline?[15] By the second half of the century many others scientists, first among them the astronomer Adolph Hirsch and the ophthalmologist Rudolf Schelske, tried to eke graphic images of nerve impulses from creatures other than amphibians.[16]

In 1868 the preeminent Dutch physiologist and ophthalmologist Franciscus Cornelis Donders took these investigations to a new level.[17] He became known as the first person to measure the duration of mental acts along with his student Johan Jacob de Jaager.[18] Donders and his student offered a number for the speed of thought—amounting to approximately a tenth of a second—and claimed for themselves the honor of having been the first to measure it. They criticized previous investigators, such as Johannes Müller, for viewing "the time within which a message is transmitted from the periphery to the spinal chord and the brain or from there to the muscles as infinitesimal."[19] In a famous publication provocatively titled "On the Speed of Mental Processes" (1868), Donders argued that "the mental process of conception and expression of the will lasts less than 1/10 of a second."[20]

Donders's work soon became known in France through "journals and scientific publications" that "have recently recounted the presentation of M. Donders of the Utrecht Academy of two extremely interesting instruments."[21]

15. The letter, written by the German theologian, politician and scientific writer Otto Ule, was sent to the geologist Eduard Desor, friend of the renowned naturalist Louis Agassiz and a prime force in establishing Hirsch's Observatory and Hipp's telegraph and electric clock factory. Ibid. "What prevents us from developing the organ of our intelligence through sound and sustained exercise? Why not aspire to virtuosity in the art of thought, from the moment it is demonstrated that it does not depend only on individual capacities, but that it can be the product of education and exercise?" Ibid., 202.

16. For these early attempts see Jimena Canales, "Exit the Frog, Enter the Human: Physiology and Experimental Psychology in Nineteenth-Century Astronomy," *British Journal for the History of Science* 34 (June 2001).

17. F. C. Donders, "La Vitesse des actes psychiques," *Archives Néerlandaises* 3 (1868). For recent work on Donders, see Henning Schmidgen, "The Donders Machine," *Configurations* 13 (2005): 211–56.

18. While Helmholtz had lamented that reaction time was related to the speed of thought, they found this relation extraordinarily interesting. De Jaager cited Helmholtz's claim that reaction time measurements "suffer from the unfortunate fact that part of the measured time depends on mental processes." Cited in Johan Jacob de Jaager, "Introduction," in *Origins of Psychometry: Johan Jacob de Jaager, Student of F. C. Donders on Reaction Time and Mental Processes (1865),* ed. Josef Brozek and Maarten S. Sibinga, Dutch Classics in the History of Science (Nieuwkoop: B. de Graaf, 1970), 16.

19. Johan Jacob de Jaager, "Reaction Time and Mental Processes," in *Origins of Psychometry: Johan Jacob de Jaager, Student of F.C. Donders on Reaction Time and Mental Processes (1865),* ed. Josef Brozek and Maarten S. Sibinga (Nieuwkoop: B. de Graaf, 1970), 43.

20. F. C. Donders, "On the Speed of Mental Processes," in *Attention and Performance II: Proceedings of the Donders Centenary Symposium on Reaction Time,* ed. W. G. Koster (Amsterdam: North Holland, 1969), 418.

21. Ramon de la Sagra, "Académie des sciences: Séance du lundi 1er février," *Les Mondes: Revue hebdomadaire des sciences et de leurs applications aux arts et à l'industrie par M. l'abbé Moigno* 19 (1869): 213. Donders's work was reprinted in F. C. Donders, "Deux instruments pour la mesure du temps nécessaire pour les actes psychiques (Extrait des Archives Néerlandaises)," *Journal de l'anatomie et de la physiologie normales et pathologiques de l'homme et des animaux* 5 (1868).

A few years after his investigations, the German psychologist Wilhelm Wundt instituted reaction time research in his Leipzig laboratory. The "standard account" of the history of reaction time and the personal equation designated Wundt as the founder of experimental psychology and described it as a characteristically German discipline.[22] To this day, his reaction time experiments are still widely considered the *crucial experiments* demonstrating that mental processes are measurable and quantifiable.

The second half of the nineteenth century was marked by a burst of new research in these topics. Personal equation experiments in astronomy were accompanied by analogous investigations by physiologists and psychologists. Many scientists in France and elsewhere publicized numbers for the speed of nerve transmission not only in animals, but also in humans. Their investigations expanded from studying motor nerves in severed body parts to studying sensory nerves in living subjects. They switched from measuring the speed of sensory transmission to determining the duration of mental acts. Various instruments came into use: Pouillet's chronoscope; Helmholtz's rotating drums; Arago's chronometers (the *chronomètre à pointage*); Perrelet's *chronomètre à détente;* Wheatstone's chronoscope; Schelske's Krille registration apparatus; Hipp's chronoscope; Donders's noematachometer and noematachograph; Marey's drums; Henkel's apparatus; de Jaager and Donders's phonautograph; and the astronomers' artificial transit machines. In the span of a few years, reaction time experiments shifted from being largely criticized by the scientific community to becoming foundational for a new discipline.

Although there were important differences in instrumentation, the overall conception of the experiment remained the same. When a subject reacted to a stimulus, the lag time between stimulus and response was recorded. Scientists separated the duration of thought from the speed of sensory transmission by using "the method of differential nerve lengths" based

22. For a history of the continued association of experimental psychology with reaction time experiments, Germany, and Wundt, see Mitchell G. Ash, "The Self-Presentation of a Discipline: History of Psychology in the United States between Pedagogy and Scholarship," in *Functions and Uses of Disciplinary Histories,* ed. Loren Graham, Wolf Lepenies, and Peter Weingart (Dordrecht: D. Reidel Publishing Co., 1983). For accounts that consider French experimental psychology as limited and flawed compared to German, see Françoise Parot, "La Psychologie scientifique française et ses instruments au début du XXe siècle," in *Studies in the History of Scientific Instruments,* ed. Christine Blondel et al. (London: Rogers Turner Books, 1989); Kurt Danziger, *Constructing the Subject: Historical Origins of Psychological Research,* ed. William R. Woodward and Mitchell G. Ash, Cambridge Studies in the History of Psychology (Cambridge: Cambridge University Press, 1990); John I. Brooks III, "Philosophy and Psychology at the Sorbonne, 1885–1913," *Journal of the History of the Behavioral Sciences* 29 (April 1993). For an account of the continuing failure of experimental psychology in France after the death of Théodule Ribot, see Serge Nicolas and Ludovic Ferrand, "Pierre Janet au Collège de France," *Psychologie et histoire* 1 (2000).

on changing the place in the body to which the stimulus was applied. The speed of nervous transmission was subtracted from the total reaction time to reveal the sought after *speed of thought.*

Mental Origins

The association of these tenth-of-a-second delays with measurements on the speed of thought, however, only appeared later, slowly and polemically. In France, this interpretation first gained currency with a widely reprinted article provocatively titled "La Vitesse de la volonté" (1867), written by the scientist and popular science writer Rodolphe Radau.[23] Astronomers, Radau claimed, had measured the speed of thought: "Thought is not born instantaneously. It is a natural phenomenon subject to the laws of time and space. The lost time is not the same in different observers: one perceives, reflects and moves faster than the other. . . . This explains the differences which have been repeatedly noticed by astronomers who have observed the same phenomenon. Two people have never seen a star pass at the same time behind a vertical wire." Their discovery had, grandiosely, turned the "problem of life" into an "exact science."[24]

Insisting that reaction time experiments measured the time of volition and thought, Radau ignored contrary interpretations favored by other astronomers and physiologists (of which he knew well). He drew heavily on du Bois-Reymond's lectures on "the transmission speed of volition and sensation through the nerves" that were translated in the popular *Revue scientifique.* In them, du Bois-Reymond argued that measurements of "the speed of voluntary and sensory transmission" were a continuation of the work by the eighteenth-century physiologist Albrecht von Haller, who measured the

23. This interpretation contrasted with the one presented in a previous article on the problem of personal errors in astronomy. In it Radau described various interpretations of reaction time. He documented how a number of scientists believed that in well-trained observers the personal equation was a small, immeasurable, relatively uninteresting, and purely physiological phenomenon, equal or analogous to the time of visual persistence. Radau, initially, did not consider reaction time experiments as unambiguously measuring the speed of thought. By his second piece, his initial caution disappeared. His first article was Rodolphe Radau, "Sur les erreurs personnelles," *Le Moniteur scientifique* 7 (1865), and "Sur les erreurs personnelles," *Le Moniteur scientifique* 8 (1866). It appeared in German in Rodolphe Radau, "Ueber die persönlichen Gleichungen bei Beobachtungen derselben Erscheinungen durch verschiedene Beobachter," *Repertorium für physikalische Technik für mathematische und astronomische Instrumentenkunde, herausgegeben von Dr. Ph. Carl* 1 (1865), and "Ueber die persönliche Gleichungen," *Repertorium für physikalische Technik für mathematische und astronomische Instrumentenkunde, herausgegeben von Dr. Ph. Carl* 2 (1866). These works were cited by Exner and Wundt. Radau used the following sources for his account of Maskelyne: *Zeitschrift für Astronomie, herausgegeben von Lindenau und Bohnenberg* 2 (1816). For Bessel he used C. A. F. Peters, *Ueber die Bestimmung des Längenunterschiedes zwischen Altona und Schwerin, ausgeführt im Jahre 1858 durch galvanische Signale* (Altona: Hammerich & Lesser, 1861).

24. Rodolphe Radau, "La Vitesse de la volonté," *Le Moniteur scientifique* 5 (1868): 91.

time taken by different individuals to read Virgil's *Aeneid* out loud and as fast as possible.[25] In an uncanny coincidence, the speed of nervous transmission found by Haller was approximately the same as the one found almost a century later by du Bois-Reymond's friend and colleague, Helmholtz.[26] Du Bois-Reymond and Radau argued that the decisive step in transforming these experiments into measurements for the speed of thought was currently undertaken by Donders.

Étienne-Jules Marey, the physiologist best known for his use of the graphic method and chronophotography, also adopted this particular interpretation. He became heavily invested in interpreting reaction time experiments as mental, providing an important justification for his earlier frog-muscle investigations. In his opening lecture at the Collège de France for his course on "nervous and mental speed," he framed reaction time as the speed of thought. According to Marey, Donders's recent revelations arose from the comparatively less exciting experiments on frogs he had pioneered. He argued that early experiments on frogs had led to experiments on mental acts. The "knowledge of nervous action which became possible from previous studies on muscular acts," he explained "permitted, in turn, their elevation to the study of mental acts."[27] For years the association of Marey's work on frogs with measurements on the speed of human thought continued largely through his own initiative and that of his supporters. In a tribute to Marey's work, for example, the graphic method was described as central to Donders's work: "Ask M. Donders how the graphic method permitted to separate this physiological time into two parts: the speed of transmission of sensation and volition through the . . . nerves, and the duration of the cerebral work of perception, psychic time."[28] These associations were supported by Marey himself, who among the applications of his graphic method included the "astronomers' personal equation."[29] But at the time of Marey's publication, and for more than a decade already, astronomers were not measuring the personal equation with Marey's methods, but rather

25. The use of the term "reader" to describe the experimenter persisted well into the late nineteenth century.

26. Emil du Bois-Reymond, "Vitesse de la transmission de la volonté et de la sensation à travers les nerfs: Conférence de M. du Bois-Reymond à l'Institution Royale de la Grande-Bretagne," *Revue scientifique* 4, no. 3 (15 December 1866).

27. Etienne-Jules Marey, "Leçon d'ouverture: Vitesse des actes nerveux et cérébraux.—Le Vol dans la série animale, Collège de France, histoire naturelle des corps organisés, cours de M. Marey," *Revue scientifique* 6, no. 4 (26 December 1868): 62.

28. Gavarret, "Observations à l'occasion du procès-verbal. III. Méthode graphique," *Bulletin de l'Académie de médecine* 7 (1878): 760–61.

29. Etienne-Jules Marey, "La Méthode graphique dans les sciences expérimentales," *Travaux du laboratoire de M. Marey* 1 (1876): 145.

with their own artificial transit machines.[30] Furthermore, the interpretation given to the personal equation as mental and its history as originating in 1796 was fiercely debated within various scientific circles. This "standard account" was only one of many others.

Longitude and Astronomy

Essential elements of the "standard account" already appear in some of the earliest descriptions of the discovery of the personal equation. Helmholtz, for example, mentioned the "remarkable fact discovered by Bessel" and frequently referred to the work of astronomers.[31] A few years later the French scientist François Arago, known for introducing the daguerreotype to the general public, gave a much more detailed history of the problem. He explained how Maskelyne, the Astronomer Royal at Greenwich, first noticed individual differences at the end of the eighteenth century.[32] His interest in the problem's history, however, was largely driven by current concerns, where the tenth of a second appeared prominently.

Arago explained how in one particular area of astronomy attention to the tenth of a second was not a "vain luxury," but an urgent need. He referred to the determination of time and longitude where astronomers noted the time in terms of "hours, minutes, seconds, and even tenths of a second." Since "a tenth of a second in time is equivalent to nothing less than a second and a half of arc," ten accumulated mistakes of this magnitude could result in map discrepancies of nearly half a kilometer.[33]

Arago grew increasingly preoccupied with the tenth of a second. In 1853, a few months before his death, he remarked on the need for new chronometers with which he "could read without any doubt tenths of a second."[34] He marveled at a rare instrument now "in my possession, made in Vienna" in which the hand, instead of making a complete turn in a minute, "made a complete turn in a second." He also called attention to new American-built

---

30. Marey went to the Observatoire de Paris to learn about Wolf's machine for measuring the personal equation. Marey, *La Méthode graphique dans les sciences expérimentales et principalement en physiologie et en médecine*, 148. Marey also cited the work of the astronomers Prazmowski, Hänckel, Hirsch, and Plantamour.

31. Helmholtz, "Ueber die Methoden, kleinste Zeittheile zu messen, und ihre Anwendung für physiologische Zwecke," 325, and "On the Methods of Measuring Very Small Portions of Time, and Their Application to Physiological Purposes," 189.

32. François Arago, "Note sur un moyen très-simple de s'affranchir des erreurs personnelles dans les observations des passages au méridien," *Comptes rendus des séances de l'Académie des sciences* 36 (1853).

33. These errors were larger the closer observations were to the equator. Ibid.

34. Ibid.

electric clocks that could be adapted for measuring these short periods of time.[35]

Arago's interest in the exact determination of longitude was not only due to its importance for mapmaking and navigation. It was also relevant for the precise determination of the length of the meter. He claimed that "many years ago" errors of this magnitude had been used by a member of the Académie as an explanation of the "extraordinary and difficult to explain anomaly" that haunted the famous meter bar of 1799.[36] That meter bar had been defined as a 1/10,000,000th part of the earth's quadrant circumference, and errors in its determination spread to those of the meter based on it.[37]

Joseph Delambre and Pierre Méchain were in charge of a vast project to determine the length of the meter. Their main task was to provide the geodesic measurements for this project, which led them to embark on a seven-year-long odyssey considered to be the most important scientific mission of the era. Yet when Napoleon unveiled a new platinum meter bar based on their results, this standard faced a difficult future. Although astronomers who participated in the project presented an optimistic picture to the public, they knew it housed a dangerous error.

Tenth-of-a-second errors in determinations of the earth's circumference appeared at the center of a crisis—a crisis striking at the essence of measurement. These same errors haunted Arago's other projects. They were the "only key" that he believed might account for the "paradoxical and constant variations" found by Alexander von Humboldt and himself when they determined the latitude of Paris in 1809. Nearing the end of his life, Arago dusted off various essays published as early as 1816 and proudly proclaimed that he had studied personal equation errors many years before a furor of interest started surrounding them.[38] He reminded his colleagues of an 1842 publication where he had brought to the attention of the Académie "a cause of error which up to now had never appeared in an analogous work" affect-

---

35. Ibid.

36. François Arago, "Rapport sur deux mémoires présentés, l'un par M. Eugène Bouvard, l'autre par M. Victor Mauvais, relatifs à l'obliquité de l'écliptique," *Comptes rendus des séances de l'Académie des sciences* 15 (1842): 946, and "Mémoire sur les cercles répétiteurs," in *Oeuvres de François Arago* (Paris: Théodore Morgand, 1865), 121.

37. For the history of the expedition to find the length of the meter see Ken Alder, *The Measure of All Things: The Seven-Year Odyssey and Hidden Error That Transformed the World* (New York: Free Press, 2002). Alder mentions the relation to the personal equation on p. 307.

38. This work was first published in 1816 in the *Connaissance des temps* for 1813. Arago remarked on errors that only surfaced when different observers were involved. Tests on the accuracy of instruments, he suggested, should always be done "under the same circumstances, and especially by the same person." Arago explained how the source of these errors might lay in a "visual defect" of the observer. He called them "erreurs constantes de pointé." Arago, "Mémoire sur les cercles répétiteurs," 128.

ing two of his junior astronomers.[39] He referred to the contested concept of a "personal," "distinct," and "individual collimation" that "varies from one observer to the other, and for the same observer, according to the eye used."[40] In his early work Arago had insisted that these "distressing and so singular anomalies" were most probably due to optical and physical causes, but the problem of individual biases soon became more complicated.[41] He and others started to look beyond optics.

Personal equation errors plagued the traditional skill-intensive "eye-and-ear method" for determining longitude, which was difficult, requiring the full attention of highly trained and talented astronomers. This method consisted in estimating the position of a star across the reticules of a telescope between successive one-second-long clock beats. Talented observers could determine the star's position at tenth-of-a-second intervals, but errors were frequently larger. Arago argued that if astronomers used chronometers to note the time of a star's passage (instead of estimating this moment between one-second clock beats), these errors might be significantly reduced.[42]

Starting in the 1860s astronomers increasingly used telegraphic methods (sometimes referred to as the "American method" or "electro-chronography") for time and longitude determinations.[43] These technologies permitted them to determine time by pressing a key when a transit star passed through the reticules of their telescopes and recording the moment on a moving strip of paper. With these instruments, they replaced the "eye-and-ear method." Yet even with telegraphy and automatic inscription devices, the best observers were rarely able to avoid errors of the order of a tenth of a second. While working with these electric technologies, scientists became

39. François Arago, "Rapport sur deux mémoires présentés, l'un par M. Eugène Bouvard, l'autre par M. Victor Mauvais, relatifs à l'obliquité de l'écliptique," 945–46.

40. Ibid., 945.

41. Arago warned how they could change according to the position of the body of the observer vis-à-vis the stars and also with respect to the eye used. Ibid., 946.

42. Arago believed that tenth-of-a-second errors arose in connection to mental work, and he prompted astronomers all over the world to adapt their observational methods to reduce mental fatigue. Astronomers became part of the larger effort to mechanize work and mental processes, concerning industrialists, accountants, and mathematicians like Charles Babbage, the creator of some early calculating machines. In his personal notebook, Babbage explained how fatigue affected his own personal equation. "There was however not merely a personal equation but a *periodic* personal equation. For I found on trial that my own accuracy . . . varied with the state of bodily fatigue." Charles Babbage, Papers on Astronomy, 1862, British Museum Library Manuscript Collection (emphasis in original). These observations were alluded to much earlier, but without attributing them to the "personal equation," in Charles Babbage, *Reflections on the Decline of Science in England, and Some of its Causes* (New York: A. M. Kelley, 1830; reprint 1970), 173–74.

43. The Bonds at Harvard and Adolph Hirsch of Neuchâtel were some of the first to use these technologies. The Bonds used an electric clock, similar to the one used by Alexander Dallas Bache for the trigonometric drawing-up of the U.S. coastline.

increasingly concerned with the time that elapsed between stimulus and response. They inquired into the speed of electric transmission, the inertia of inscription devices, and lag times due to observers.[44]

Shortly before his death, nearly blind, and a disillusioned witness to yet another hijacked republic, Arago portrayed the most important years of his career in a manner that was dramatically different from that which he had previously revealed. He talked about observation errors in public and loudly, freely reminiscing about the role of "observational errors" in his decades-old investigations.[45] Arago remembered how at the time, a number of his colleagues did not accept his theory. Most prominently among them was the German mathematician Friedrich Gauss, who was responsible for developing an alternative theory of errors and who examined Arago's work "with some severity in the Gazette littéraire of Leipzig."[46]

Gauss developed a mathematical technique for eliminating errors, frequently called the "least squares" method, which consisted in giving observations lesser weight depending on the square of their deviation from the mean. His work enabled investigators to sort errors that would eventually be called "random" or "accidental" from those called "constant" or "systematic." The pattern of the former would follow the famous bell-shaped curve, which was also called "Gaussian."

In contrast to Gauss, Arago considered observation errors of an entirely different nature. These persisted even when observations were repeated and averaged and even when subjected to the least squares method. They did not display the same bell-curve shape as random errors, and they did not disappear by using traditional instruments (such as microscopes and telescopes) previously legitimated as expanding the reach of the senses. These errors could no longer be explained away by invoking the well-known universal fallibility of the senses or by using these new mathematical techniques.

In the Paris Observatory, research on individual differences in observation started in the 1840s under Arago and increased dramatically in the

---

44. Another complication involved understanding how the speed of nerve transmission was affected inertially. How were reactions to stimuli affected by the movement of the observer, for example, when traveling on a train? Scientists needed to account for acceleration effects on the nervous system: "Now suppose that the mechanic on a locomotive of an express train traveling one mile (English) per minute extends his arm towards the tender and moves the fingers, then the movement of the nervous agent is destroyed by the movement of the train." This situation was "the same as when the movement of a cannon ball thrown towards the equator from the west is destroyed by the movement of the earth around its axis. In that case it is not the cannonball that hits the boat, but rather the boat that hits the cannonball." Bois-Reymond, "Vitesse de la transmission de la volonté et de la sensation à travers les nerfs: Conférence de M. du Bois-Reymond à l'Institution royale de la Grande-Bretagne," 39.
45. François Arago, "Sur les observations des longitudes et des latitudes géodésiques," in *Oeuvres de François Arago* (Paris: Théodore Morgand, 1865), 147.
46. Arago, "Mémoire sur les cercles répétiteurs," 120.

Corrections personnelles à appliquer aux heures de passage observées de 1837
à 1853.

	ED.	L.	P.	VM.	G.	F.	IV.	B.	CM.	LL.
1837.	0ˢ,00	−0ˢ,01	+0ˢ,44	0ˢ,00						
1838.	0,00	−0,04		0,00						
1839.	+0,02	−0,07		0,00						
1840.	+0,04	−0,10		0,00						
1841.	+0,05	−0,13		0,00	−0ˢ,19					
1842.	+0,05	−0,13		0,00	−0,38					
1843.	+0,05	−0,13		0,00	−0,50	+0ˢ,01				
1844.	+0,06	−0,13		+0,01	−0,53	+0,05				
1845.	+0,06	−0,13		+0,02	−0,57	+0,09	0ˢ,00	−0ˢ,12		
1846.	+0,06	−0,13		+0,03	−0,61	+0,14	0,00	−0,12		
1847.		−0,13		+0,04	−0,65	+0,19	0,00	−0,12		
1848.		−0,14		+0,04	−0,61	+0,15	−0,02	−0,09		
1849.		−0,14		+0,05	−0,59	+0,12	−0,02	−0,08		
1850.		−0,14		+0,07	−0,58	+0,10	−0,02	−0,07		
1851.		−0,13		+0,09	−0,56	+0,07	−0,03	−0,06	−0ˢ,35	
1852.		−0,13		+0,12	−0,54		−0,03	−0,06	−0,40	−0ˢ,14
1853.		−0,15		+0,16	−0,52		−0,04	−0,06	−0,43	+0,04

Figure 2.2. Personal corrections for observers published under the directorship of Le Verrier applied to observations from 1837 to 1853. The initials above each column designate the observers. From "Réduction des observations faites aux instruments méridiens," *Annales de l'Observatoire impérial de Paris, Observations* 2 (1859), p. xli.

decades after his death during the Second Empire. When the astronomer Urbain Le Verrier succeeded Arago as director of the Paris Observatory, the problem was so pervasive that he decided to apply personal corrections retrospectively: to the observations done under Arago from 1837 to 1853. Every observation was shifted in time by a few tenths of a second, forward or backward, depending on the observer. In the 1850s observers at most major astronomical observatories started to be tested for their personal equations. By the end of Napoleon III's reign in the 1870s, all the observers in the Paris Observatory were tested.[47]

The astronomer Hervé Faye was particularly pessimistic about the extent of these errors. All "determinations of absolute time" used in astronomy, Faye noted, were implicated: "One knows, since the beginning of this century, that this determination [of time] is completely illusory." According to him, only "relative time" could be found, and this one, ironically, with a much greater precision. Methods for determining time and longitude—including telegraphic ones—had to contend with the problem that "sensations separated by a very real interval will be falsely noted as simultaneous."[48]

47. François Gonnessiat, *Recherches sur l'équation personnelle dans les observations astronomiques de passage*, Annales de l'université de Lyon (Paris: G. Masson, 1892), 3:159.
48. Hervé Faye, "Sur les erreurs d'origine physiologique," *Comptes rendus des séances de l'Académie des sciences* 59 (12 September 1864): 475.

The problem plagued not only France and its observatories, but "all places where one observes, with an admirable precision, astronomical phenomena." The scourge reached as far as Poulkova, Koenigsberg, and Greenwich, sites of the most important observatories and, places where "one could not find two observers who would reach accord on absolute time."[49]

Because of the problems he noticed in astronomical practices, Faye was led to the conclusion that the human "spirit" was not infinitely fast:

> Imagine that the spirit is an eye in the center of the brain; an eye attentive to the changes that each sensation causes in the nerve cells that lead to it. If similar sensations occur at the same place, this internal eye will easily tell whether they occurred at the same time or one after another. But if these sensations come from different senses where the nerves end up at different parts of the brain, the internal eye will need to move in order to go from one area to another, and the time spent doing this will not be detected. Sensations separated by a very real interval will be wrongly perceived as having occurred at the same time. The lost time used to go from one sensation to the other can amount to over a second. It will in fact vary from one individual to another, depending on the speed at which the internal eye moves to successively study the keys of this prodigiously complex keyboard [clavier] that is called the brain.[50]

The famous physicist Henri Victor Regnault (whom Helmholtz described as the best French scientist) explained how determining these errors was urgent. The "non-instantaneity of transmission, not only in personal organic sensations, but also in the telegraphic registering apparatus" was a matter about which "one must worry enormously." Yes, it was important to determine the time of transmission in the wires of the equipment and the inertia

49. Hervé Faye, "Sur les observations du soleil," *Comptes rendus des séances de l'Académie des sciences* 28 (1849): 243.

50. Individual differences in observation, according to Faye, were caused by the different speeds of the "internal eye." For Faye a "lost time" arose when different impressions were compared to each other. This delay was mostly due to the mind, since Faye surmised that nerve transmissions "probably" occurred at a "rapidité toute électrique" and "should be almost the same for all individuals." This mental work was fatiguing: "The fact remains that the need to compare two sensations coming from different origins forces the spirit [*esprit*] to a very peculiar task, as it spends such a considerable amount of time establishing communication between different nerve cells. Moreover this task is very tiring, while the comparison of sensations from the same origin is not, or much less so." Faye, "Sur les erreurs d'origine physiologique," 475–76. Cited in Gonnessiat, *Recherches sur l'équation personnelle dans les observations astronomiques de passage*, 120, and translated in Sanford, "Personal Equation," 413 n. 1. Even the auditory method of coincidences long used for comparing and synchronizing clock signals was affected by these personal errors. Although the trained ear registered coincidences in rhythmic sounds to an uncanny degree, this method was not free from them. Faye impressed Airy in 1854 by noticing a difference between two pendulums separated by four meters using the method of coincidences. Hervé Faye, "Sur la méthode des coïncidences appliquée à la mesure de la vitesse du son et sur la détermination des longitudes," *Comptes rendus des séances de l'Académie des sciences* 55 (1862).

of the registering apparatus, but Regnault insisted especially on "the errors due to the personal appreciation of the observer," which were "significantly larger."[51] Scientists like Regnault noticed that too many experiments and measurements were predicated on an architecture of the body that ignored the time of internal transmissions and individual differences. New investigations on personal errors questioned the legitimacy of these experimental systems. If time and length were not absolute, what was?

Measurements of Lengths

Another important aspect of the personal equation that was obscured in standard accounts pertained to other, even simpler, measurements, such as length. Cutting across the life and physical sciences, personal equation errors affected disciplines far from astronomy—everywhere exact measurements were used. The history behind these efforts can help us understand how measurement gained the privileged position it would hold for the next century. It can help us voice the complex negotiations behind the establishment of measurement-based science.

In most accounts, measurement appears as a simple, straightforward activity. Inasmuch as it is simple, it is not worthy of examination. Yet numerous difficulties lay behind the establishment of measurement-based science. These measurement problems were interdisciplinary. Challenges faced by astronomers and physicists were the same as those affecting anthropometric measurements used to identify individuals for policing and colonial purposes.[52]

Scientists noticed that in the process of measuring the size of rulers, different individuals assigned slightly different lengths to these standards. These errors, referred to as "individual collimations" or "erreurs de pointé," appeared when leveling an instrument or, when using a micrometer, when bringing two marks in line with each other. They also surfaced when observers tried to find the exact center of circles and dots and the precise alignment between distinct objects—whenever scientists set the crosswires of an instrument on the division mark of a scale. Personal equations soon appeared even in such apparently simple activities as measuring the diameter

---

51. These "were always significantly larger than those which come from telegraphic register systems." Regnault, "Remarque de M. Regnault à l'occasion de la note de M. Faye," *Comptes rendus des séances de l'Académie des sciences* 59 (1864): 479–80.

52. The criminologist Alexandre Lacassagne remarked on the "serious obstacle of the personal equation." Alexandre Lacassagne, *Alphonse Bertillon: l'homme, le savant, la pensée philosophique* (Lyon: A. Rey, 1914), 16.

of the sun and moon.[53] Metrologists concerned with determining standards of length were forced to measure the personal equation of their observers. At times, they even asked observers to sign their observations. These errors constituted a particular danger since they affected both spatial and temporal measurements. They spread to affect perceptions of brightness, taste, temperature, and weight.

Scientists traced the spread of personal equation problems to measurements of length to the comparison of English yards undertaken by the English astronomer Francis Baily in 1835.[54] That year Baily undertook a detailed project to measure and determine the length of different standards. In the process, he noticed a problem that arose when he tried to bisect the dots of a ruler with the cross wires of his measuring instrument. Because these dots resembled stars, he explained, his measurements of length were affected by the astronomical problem of the personal equation. Baily's work was alarming since it showed that the problem noticed by astronomers when reacting to a transit star crossing the wires of a sighting device was also present in passive, leisurely observations: "I believe that it seldom happens that two persons, chosen indifferently, will agree precisely in their measures of a line, or a dot with the cross wire of the micrometer; but that there will almost always be some slight difference between their results: similar to what takes place in observations with transit instruments, where this anomaly is styled the *personal equation,* a term which I shall employ on the present occasion."[55]

Baily vacillated between thinking that individual differences were due "to a peculiar state of vision, or to a peculiar mode of making the bisections,"

53. A personal equation in the observation of the borders of the sun and moon was frequently noted. In the Paris Observatory investigations on these differences were carried out in 1856 and published in the *Annales de l'Observatoire de Paris, observations* 12 (1860): 131. They were described in Gonnessiat, *Recherches sur l'équation personnelle dans les observations astronomiques de passage,* 88. Le Verrier remarked on the individual differences in estimating the diameter of the sun in 1858. He attributed these errors to "systematic differences in the manner of observing." Urbain Le Verrier, "Théorie et tables du mouvement apparent du soleil," *Annales de l'Observatoire impérial de Paris* 4 (1858). Encke's number for the sun's diameter (31′56.84″, deduced from the eighteenth-century transits of Venus) differed from many others, especially from the one Le Verrier had found by observing transits of Mercury (32′0.02″). Both of these values differed markedly from the ones obtained from direct measurements during an 1842 total eclipse.

54. Arago disagreed with the attribution of the "origin" of the discovery of personal errors in measurements of length to Baily. He called it a *petite malice* from Faye. Arago insisted it was him, and not the British astronomer, who was the discoverer. "The *petite malice* that the author [Faye] permitted himself could have been very tasteful if the year 1816, date of the publication in the *Connaissance du temps* of a frequently cited Mémoire, had not preceded the year 1834 or 1835." Arago, "Sur les observations des longitudes et des latitudes géodésiques," 147. This work summarized numerous discussions that took place at the Académie des sciences during the early part of 1853.

55. Francis Baily, "Report on the New Standard Scale of this Society. Drawn Up at the Request of the Council, by F. Baily, Esq. F. R. S. and C., and One of the Vice-Presidents of the Society. Presented December 11, 1835," *Memoirs of the Royal Astronomical Society* 9 (1836): 92. Italics mine.

1834.	N° of Obs.	Scales.	Person. Equat.	Mean.	Observers.
			div.	div.	
April 5	10	Ast. Soc.	0·19	0·19	Murphy and Baily.
April 2	10	Shuckburgh	6·64	6·64	Murphy and Donkin.
April 3	10	Russian	7·19	} 7·04	Murphy and Donkin jun.
April 4	10	Ast. Soc.	6·89		
March 22	5	Ast. Soc.	4·80		
March 24	20	Danish	5·70		
April 3	10	Shuckburgh	5·76	} 4·72	Murphy and Henderson.
April 4	10	Russian	1·47		
April 5	10	Ast. Soc.	4·96		
March 24	8	} Danish {	3·27	} 4·08	Murphy and Johnson.
March 25	10		4·89		
March 24	10	Danish	4·40	4·40	Baily and Donkin.
April 4	10	Ast. Soc.	3·08	3·08	Baily and Henderson.
April 3	10	Russian	4·30	} 3·07	Henderson and Donkin jun.
April 5	10	Ast. Soc.	1·84		
April 3	10	Shuckburgh	1·36	1·36	Johnson and Henderson.

Figure 2.3. Table comparing the length of different standards of measurement. The middle column shows the observers' personal equation when measuring different scales of length. From Francis Baily, "Report on the new Standard Scale of this Society. Drawn up at the request of the Council, by F. Baily, Esq. F. R. S. and C., and one of the Vice-Presidents of the Society. Presented December 11, 1835," *Memoirs of the Royal Astronomical Society* 9 (1836): 35–184, on p. 110.

but something was clear: "Each individual has some real or imaginary cause of preference for selecting the precise portion of the line or dot under consideration, which may differ from that other person."[56] Part of the problem arose because rulers' dots were frequently "not circular, but of an irregular pear-shaped form." Baily suggested using the short, straight lines used today instead of dividing the standard with "enormous and irregular" dots.[57] He contended with fatigue, increased the number of observations, measured the personal equation of different observers, and had them exchange places from one side of the ruler to the other. He fought against the "distracted attention of the observer" by employing an amanuensis.[58]

His investigations showed inherent difficulties in measuring, and even worse, in making "true copies." Even the simplest measuring tasks required

56. Ibid., 96.
57. Ibid., 92. Baily contrasted these dots to the "fine" and "scarcely perceptible" lines of the French meter.
58. Ibid.

complicated and highly disciplined bodily skills.[59] The standard should remain "untouched by the hands of any clumsy or inexperienced workman, or experimentalist"[60] and only be handled by those "conversant with micrometrical measurements."[61] Anything that touched a ruler would alter its pristine length. Focus now switched to the ruler's bedding and to things affecting the bedding—not only temperature but also the floors and walls of the laboratory. These concerns led him to an infinite regress that spread from the inside of the observer to the world outside of the laboratory. But the problem did not end there. In despair, he recommended that all measurements involving rulers, even those undertaken with a micrometer, be signed.

The endeavor, at times, seemed self-defeating. How could scientists escape the vicious circle in which both observations and reactions were affected? Astronomers altered their methods and instruments often following solutions volunteered years earlier by Baily. They isolated their workplaces. When measuring longitudes they exchanged places (say London and Paris) in the same way Baily switched observers from one end to another end of a ruler. They tried minimizing fatigue by increasing the division of labor in the observatory. They cared about the attention of observers, making sure their laboratories and observatories were silent and free from distractions.[62] Distorted, seemingly asymmetrical stars and planets were treated much like the pear-shaped dots that often divided rulers, and, using prisms, were corrected.

Baily's instructions were carefully followed by later generations. More than half a century after they were published, a prominent metrologist could still claim that Baily "gave instructions that, on many points, can serve as a guide to the metrologists of today."[63] Yet the problem did not disappear.

Beyond Astronomy and Experimental Psychology

Scientists, philosophers, and writers across disciplines and nations used the terms "reaction time" and "personal equation" outside of the institutional

---

59. Ibid., 36, 39n. For the role of bodily skills in the production of knowledge, see H. Otto Sibum, "Les gestes de la mesure: Joule, les pratiques de la brasserie et la science," *Annales HSS* 4–5 (1998).

60. Baily, "Report on the New Standard Scale of this Society. Drawn Up at the Request of the Council, by F. Baily, Esq. F. R. S. and C., and One of the Vice-Presidents of the Society. Presented December 11, 1835," 93n.

61. Ibid., 96.

62. For experiments on the disruption brought about by the regular noises of the chronograph, see Gonnessiat, *Recherches sur l'équation personnelle dans les observations astronomiques de passage*, 118.

63. J.-René Benoît, "De la précision dans la détermination des longueurs en métrologie," in *Rapports présentés au congrès international de physique réuni à Paris en 1900*, ed. Ch.-Éd. Guillaume and L. Poincaré (Paris: Gauthier-Villars, 1900), 61, 62.

confines of experimental psychology and astronomy. A famous American psychologist described the personal equation as having "interesting ramifications into physiology, psychology and anthropology."[64] Repercussions were felt elsewhere as well, in art.

The term "personal equation" slowly gained currency in the broader culture. Its meaning expanded to include a broader set of personal differences that went far beyond the differences noticed by astronomers in the timing of star transits. Through the course of the century it became a term used to describe personal opinion and bias. The following definition from Webster's dictionary reveals the complex meaning of the term, ranging from astronomy to everyday judgments:

> Personal equation: The difference between an observed result and the true qualities or peculiarities in the observer; particularly the difference, in an average of a large number of observations, between the instant when an observer notes a phenomenon, as the transit of a star, and the assumed instant of its actual occurrence; or, relatively, the difference between these instants as noted by two observers. It is usually only a fraction of a second;—sometimes applied loosely to differences of judgment or method occasioned by temperamental qualities of individuals.

In a book titled *The Personal Equation* (1925), Louis Berman, a famous endocrinologist who explained how hormones affected personality, also remarked on the changing meaning of the term: "'The personal equation' is a phrase that was first invented in the eighteenth century in relation to errors in astronomical observations. . . . By transfer of the general meaning of the term the phrase came to be applied to all those individual peculiarities and idiosyncrasies which have to be taken into account in estimating personality."[65] In the 1920s, an important scholar of American literature maintained that "criticism is the science of the personal equation."[66] During this period, a whole biographical genre, titled the personal equation, was born.[67]

64. In 1888 Edmund C. Sanford, a psychologist and editor of the *American Journal of Psychology*, scrutinized the work of astronomers in a widely read history of the personal equation that inspired a generation of American experimental psychologists. Sanford, "Personal Equation," 25. He considered it "important alike to astronomy and anthropology." Ibid., 9.

65. Louis Berman, *The Personal Equation* (New York: Century Co., 1925), 299–300.

66. Percy H. Boynton, *Some Contemporary Americans: The Personal Equation in Literature* (Chicago: University of Chicago Press, 1924).

67. Harry Thurston Peck, *The Personal Equation* (New York: Harpers, 1897); Lawrence McTurnan, *The Personal Equation* (New York: Atkinson, Mentzer & Grover, 1910); Albert Guérard, *Personal Equation* (New York: W. W. Norton and Co., 1948). The Nobel Prize winner Eugene O'Neill used the term as the title for an unpublished play (1915).

From the late nineteenth to the early twentieth century, a number of thinkers and writers used the term in a broader sense. The famous sociologist Thorstein Veblen denounced the corporatization of American universities by lamenting their rejection of the valuable element of "personal equation" (1918).[68] The New England painter Charles H. Woodbury explained how it signified "individual opinion, a preference without the possibility of proof" (1919).[69] William James, professor of philosophy at Harvard University, described the "original 'personal equation' observation of [the German astronomer] Bessel," but also employed the term in its other sense.[70] When commenting on "the methods and snares of psychology," he warned how the study "of animals, savages and infants is necessarily wild work, in which the *personal equation* of the investigator has things very much its own way."[71] While the term "personal equation" was increasingly used as a label for personal opinion in literary contexts, this meaning also appeared in scientific publications.

Opinion and Testimony

The famous mathematician Karl Pearson, Francis Galton's most loyal follower, connected the personal equation to the differences in the testimony given by different individuals in general observations. Although in his work on the personal equation he mostly focused on "observations such as are daily made in the physical laboratory or the observatory"[72] and those "typical of the measurements usually made by physicists and astronomers,"[73] his

68. "Whereas it may be fairly said that the personal equation once—in the days of scholastic learning—was the central and decisive factor in the systematization of knowledge, it is equally fair to say that in later time no effort is spared to eliminate all bias of personality from the technique or the results of science or scholarship." Thorstein Veblen, *Higher Learning in America* (New Brunswick, N.J.: Transaction, 1993 [1918]), 5.

69. Charles H. Woodbury, *Painting and the Personal Equation* (Boston: Riverside Press, 1919), 57.

70. William James, *Principles of Psychology* (New York: Dover Publications, 1950), 1:413. For the broader sense in which William James referred to the personal equation, see also his "Report on Mrs. Piper's Hodgson-Control," *Proceedings of the American Society for Psychical Research* 23 (1909): 2–21, reprinted in *William James on Psychical Research*, ed. Gardner Murphy and Robert O. Ballou (Clifton, N.J.: Augustus M. Kelley, 1973).

71. James, *Principles of Psychology*, 1:194. Cited in Sonu Shamdasani, *Jung and the Making of Modern Psychology: The Dream of a Science* (Cambridge: Cambridge University Press, 2003), 34. Italics mine.

72. Karl Pearson, "On the Mathematical Theory of Errors of Judgment with Special Reference to the Personal Equation," in *Early Statistical Papers* (Cambridge: Cambridge University Press, 1948), 432. Originally published as Karl Pearson, "On the Mathematical Theory of Errors of Judgment with Special Reference to the Personal Equation," *Philosophical Transactions of the Royal Society of London* 198 (1902).

73. Pearson, "On the Mathematical Theory of Errors of Judgment with Special Reference to the Personal Equation," 379. Pearson discovered that the typical manner of calculating probable errors, consisting in using the "least squares" technique developed earlier in the century, was insufficient. In order to apply the least squares method to measurements, errors had to be random and should follow the famous bell-shaped curve. Yet Pearson's showed that "personal equation" errors were not random and that they did

conclusion informed "all types of observation." After reviewing the work of French and British astronomers, he concluded that "there is a real individuality in observation which manifests itself in the personal equation."[74] Observational errors were not entirely random, as previous researchers had often assumed, but they depended on "mental or physical likeness" and "common elements in personalities" of the observers. Neither averaging results or applying the "least squares" method solved the problem.

Pearson's investigations on the personal equation led him to rework traditional concepts of truth and testimony. In the eighteenth century, Hume argued that the testimony of individuals could never outweigh the empirical testimony of the uniform laws of nature. Hume's insight was subsequently developed into a probability theory of testimony, where accounts given by independent witnesses were considered to be mathematically independent. More than two centuries later, Pearson noticed that the testimony of different individuals could be interrelated and correlated in complex ways. His conclusions "appear to vitiate very largely the existing theory of the probability of testimony."[75]

In contrast to Hume's classic theory of testimony, Pearson took into account bodily—not only moral—differences and similarities between witnesses. His conclusion was radical: testimony was affected by biometrical "correlations" that appeared even if observers witnessed "independently." In the course of his experiments, he found that individual differences in measurements were sometimes correlated to genetic differences. Personal equation correlations could be "as high as that of a measure made on a pair of brothers."[76] After his work on the personal equation, correlation techniques, where the value of observations made by independent observers was given a probabilistic weight depending on how they related to each other, flourished.

---

not follow the normal distribution curve. Ibid., 400. "The distribution of errors of judgment can diverge in a very sensible way, both on account of asymmetry and of flat-toppedness, from the Gaussian curve of errors." Ibid., 424. In the 1890s Pearson designed a special machine to measure the personal equation of observers, and he submitted himself to it for nearly five years. The machine reflected a moving bright line of light unto a sheet of paper. The person doing the experiment was asked to mark the line's position at the ringing of a bell. The position of the line as determined by the machine and its *perceived* place were then compared to each other in order to deduce the observer's personal equation. Pearson became particularly interested in comparing the personal equation of different observers and correcting for any unforeseen correlations among them. Referring to Dr. Lee and Dr. Macdonell, two colleagues who joined him in his personal equation experiments along with Horace Darwin, Charles Darwin's son, he explained: "If Dr. Lee and Dr. Macdonell assert that a bright light was in certain position when the bell rang, their united testimony is very far from having the weight it would have on the old mathematical theory that they are independent witnesses, and yet they record perfectly 'independently.'"

74. Ibid., 433n.

75. Ibid., 433.

76. Ibid., 404.

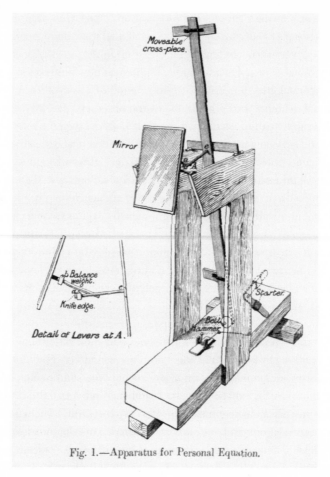

Fig. 1.—Apparatus for Personal Equation.

Figure 2.4. Personal equation machine used by Pearson. From Karl Pearson, "On the Mathematical Theory of Errors of Judgment with Special Reference to the Personal Equation," in *Early Statistical Papers*, 377–441 (Cambridge: Cambridge University Press, 1948; originally published 1902), p. 391.

Biography and the Unconscious

Personal equation and reaction time research flourished in light of its connection to other measurable elements such as race, sex, age, intelligence, and state of health. While experimental psychologists increasingly focused on reaction time measurements, other scientists found its significance to be even broader: reaction time measurements could potentially give valuable information about a person's personality and biography; they could be revealing of a person's deepest secrets or unconscious thoughts.

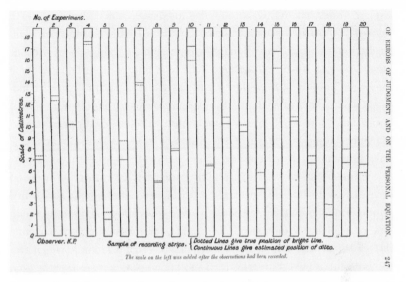

Figure 2.5. Results from Pearson's experiments showing slight disagreements between the position of a bright light, and the observer's estimation of the position. From Pearson, Karl. "On the Mathematical Theory of Errors of Judgment with Special Reference to the Personal Equation." In *Early Statistical Papers*, 377–441 (Cambridge: Cambridge University Press, 1948; originally published 1902), p. 389.

Wilhelm Ostwald, who won the Nobel Prize in Chemistry in 1909, resisted the increasing specialization of reaction time research taking place at the turn of the century in dedicated laboratories and turned to reaction time as a method for applying a "scientific point of view" to social and political problems. Many researchers, including the famous sociologist Max Weber, criticized Ostwald's "Umstulpung," or spillover, of the scientific worldview into the social sciences. Yet numerous others scientists were inspired by his attempt to broaden the reach of science beyond its clinical and laboratory moorings.

In *Grosse Männer*, published the same year he was awarded the prestigious Nobel Prize, Ostwald studied the personalities of great men, all scientists, in terms of their speed of reaction. He argued that scientific personalities could be divided into two broad categories, romantics and classics: "The speed of mental reaction is a decisive criterion for determining to which type a scientist belongs. Discoverers with rapid reactivity are romantics, those with slower reactions are classics." Reaction time, most importantly, did not have to be measured chronographically, but could be deduced from everyday engagements, such as from the time taken to respond to a student's question.[77] Ostwald reminded his readers how Helmholtz "never reacted on the instant, but only after a long time."[78] The controversial chemist did not see a particular benefit from having reaction time be short or long. Both had distinct advantages. What was important for him was how they were used to harness energy and prevent dissipation. Ostwald's biographical reaction time studies formed part of his theory of energism, an antimaterialist doctrine that stressed continuous change over physical permanence. Energy was the primary substance in the world and matter was only derivative of it.

Sigmund Freud was so intrigued by Ostwald's theory that he conceived his theory of instinct and drives in terms of similar energy flows.[79] In his famous *Psychopathology of Everyday Life,* Freud praised the reaction time studies of his favorite student Carl G. Jung, who found that reaction time tests could be used as lie detectors.[80] His method consisted in having a subject respond to a word stimulus with another word and measuring the time taken to produce a word association. Although the actual word associations became increasingly important for him, initially he was only concerned with the time of reaction. Jung noticed that "women reacted considerably more slowly" and that "uneducated subjects . . . produce much higher figures than educated ones." These initial results soon became more complicated. He investigated how reaction time varied according to the kind of word stimulus that was used: whether it was a verb, a noun, or an adjective, or whether it was abstract or concrete. He then delved even deeper by focusing on individuals, their biographies, and how these were related to their particular word associations. On one occasion a subject took an unusually long time to react to the word stimuli of "lake" and "to swim." This had an explanation: "If we ask the subject now why he hesitates at these points, we learn that once in a moment of despair he had seriously contemplated suicide by drowning." Finding that this hesitation "is quite involuntary and

77. "In what concerns applied psychology, one knows that there is a deep abyss between experimental psychology and the practical art of understanding men and managing them correctly. The results of the first, that seem easy to grasp, have accumulated an enormous mass of facts of observation and speeds of reaction, etc.; but all that . . . does not have any utility, if the goal is to understand and judge men in a general way." Wilhelm Ostwald, *Les Grands Hommes,* trans. Marcel Dufour, Bibliothèque de philosophie scientifique (Paris: Ernest Flammarion, 1912), 197.

78. Ibid.

79. The indispensability of energism was most clearly stated in his "Unconscious" (1915) publication. For energism in Freud and Lacan, see Richard Boothby, *Freud as Philosopher: Metapsychology after Lacan* (New York: Routledge, 2001).

80. Sigmund Freud, *Psychopathology of Everyday Life* (London: T. Fisher Unwin, 1914 [1901]), chap. 12, n. 8. Carl G. Jung, ed., *Diagnostische Assoziationsstudien: Beiträge zur experimentellen Psychopathologie* (Leipzig: Barth, 1906).

a kind of reflex," Jung turned to cases where he did not have "co-operation from the subject" and used this method to discover a crime.[81]

When an elderly gentleman went to Jung suspecting that his protégé, a young man of eighteen, had stolen from him, the psychologist decided to try his association experiments on the young man and quickly noticed that words such as "police," "arrest," and "jail" elicited unusually long reaction times. Finding that "the total result of this experiment appeared so convincing," he accused the man of stealing. In light of this accusation, the unveiled criminal turned "suddenly pale." After some protesting, he "burst into tears and confessed."[82]

Race

While certain writers considered the personal equation as a marker of opinion, as prejudice, and as an element affecting observations, others took it to be a privileged window into a person's mind and body. The personal equation could be a way to connect visible elements in a person (such as race) with invisible ones, such as the mind.[83] The famous British scientist Francis Galton found in the personal equation the much sought-after link between physical and mental qualities. Convinced that "the magnitude of a man's *personal equation* indicates a very fundamental peculiarity of his constitution," Galton claimed that "obvious physical characteristics" were "correlated with certain mental ones." He called on astronomical observatories to become laboratories for studying the connection between *external* appearance and *internal* constitution, recommending "that a comparison of the age, height, weight, colour of hair and eyes, and temperament . . . should be made with the amount of personal equation in each observer in the various observatories at home and abroad."[84] Research on the personal equation and reaction time was part of broader investigations on bodily and racial differences.

Richard Meade Bache, a distant relation to Benjamin Franklin, argued that the speed of reaction was inversely correlated to racial and intellectual

---

81. Carl G. Jung, "Die psychologische Diagnose des Tatbestandes," *Schweizerische Zeitschrift für Strafrecht* 18 (1905). Republished as Carl G. Jung, "The Psychological Diagnosis of Evidence," in *Experimental Researches: The Collected Works of C. G. Jung* (London: Routledge, 1973), 325.

82. Jung, "The Psychological Diagnosis of Evidence," 341.

83. For an example of the continuation of racist conclusions reached by recent investigations, see Charles Murray and Richard J. Herrnstein, *The Bell Curve: Intelligence and Class Structure in American Life* (New York: Free Press, 1994), 284. "The consistent result of many studies is that white reaction time is faster than black reaction time, but black movement time is faster that white movement time."

84. Francis Galton, "Address," *Nature* 16 (1877). For reactions to Galton's address, see Editor, "Study of Types of Character," *Mind* 2, no. 8 (1877): 573.

superiority. In "Reaction Time with Reference to Race," he noted that "inferior races" had shorter reaction times: "The popular notion that the more highly organized a human being is, the quicker ought to be the response, is true only of the higher sphere of thought, not at all of auditory, visual, or tactile impressions."[85] This was evidenced by American Indians with "wonderfully low" reaction times. His conclusion had direct implications for current events. He agreed with the controversial decision by the famous white boxer John L. Sullivan not to fight "the colored boxer, [Peter] Jackson . . . because of his race," due to how it influenced his reaction speed in the boxing ring.

The famous psychologist Jung and others used the term "personal equation" to explain the roots of competing racial and political ideologies. In his famous *Psychological Types* (1921), Jung described a personal equation of a broader psychological nature than was commonly acknowledged: "There is also a personal equation that is psychological and not merely psychophysical." This observation led Jung to distrust the "so-called objective psychology" based on "chronoscopes and tachistoscopes and suchlike 'psychological' apparatus" and to rethink notions of objectivity: "The demand that [the observer] should see *only* objectively is quite out of the question, for it is impossible."[86] The only option was to "not see *too* subjectively," and this could only be accomplished "when the observer is sufficiently informed about the nature and scope of his own personality."[87] Jung developed various methods for obtaining this information.

Jung stressed that, if not in science, in everyday life people confronted the problem of the personal equation. Although precision measurements could be obtained at the scale of wavelengths, the fact remained that in most situations we were confronted by colors—not wavelengths. The personal equation thus remained relevant because "we see colours but not wavelengths. This well-known fact must nowhere be taken to heart more seriously than in psychology. The effect of the *personal equation* begins already in the act of observation."[88] Although by the time he was writing

---

85. Meade Bache, "Reaction Time with Reference to Race," *Psychology Review* 6 (1895). Charles Richet cited this article more than two decades after it was first published. Charles Richet, "Du minimum de temps dans la réaction psycho-physiologique aux excitations visuelles et auditives," *Comptes rendus des séances de l'Académie des sciences* 163 (1916). (In the article "Bache" was misspelled as "Beach"). Richard Meade Bache was a relative of Alexander Dallas Bache who led the U.S. Coast Survey and who worked on the personal equation. The low reaction time of American Indians (lower than both of blacks and whites) was an "exception" to his theory since he determined the black race to be inferior to American Indians, while both were inferior to whites.
86. Carl G. Jung, "Psychological Types," (1921), in *The Collected Works of C. G. Jung*, ed. Herbert Read et al. (London: Routledge, 1970), vol. 6. Italics in original.
87. Ibid., 9–10.
88. Ibid., 9. Italics mine.

these lines it was no longer a problem for the exact sciences, Jung argued that it remained important everywhere else.

In 1933 Jung used the term "personal equation" to explain the anti-Semitic prejudices behind the rise of Nazism, which he was often accused of sharing. Upon becoming the editor of the famous *Zentralblatt für Psychotherapie,* he described "the differences which actually do exist between Germanic and Jewish psychology" in terms of the personal equation: "The differences which actually do exist between Germanic and Jewish psychology and which have long been known to every intelligent person are no longer to be glossed over, and this can only be beneficial to science. In psychology more than any other science there is a *"personal equation,"* disregard of which falsifies the practical and theoretical findings. At the same time I should like to state expressly that this implies no depreciation of Semitic psychology, any more than it is a depreciation of the Chinese to speak of the peculiar psychology of the Oriental."[89] While in the nineteenth century the term "personal equation" had been exclusively used to describe minute differences in the measurement results of different observers, by the years preceding WWII it represented, additionally, the source of competing ideological disagreements.

Psychometrics

Another main area of application for reaction time studies was in the growing field of psychometrics or applied psychology. These experiments were enlisted for improving production in industrial and military settings.[90] During peacetime, reaction time was studied in efforts to improve industrial efficiency. During wartime, it was used to select the best soldiers.[91]

Some of the first applications of reaction time research to industry took place at the turn of the century with Frederick Winslow Taylor. Taylor

89. Carl G. Jung, "Editorial" (1933), in *The Collected Works of C. G. Jung,* ed. Herbert Read et al. (London: Routledge, 1970), 10:533–34. In German, Carl G. Jung, *Gesammelte Werke,* 10:581. Jung's followers tried to determine Jung's own equation, and that of his mentor Freud. J. Van de Hoop, a Dutch psychiatrist, included a chapter titled "The Personal Equation" analyzing Freud's and Jung's personal equation in his *Types of Consciousness and Their Relation to Psychotherapy* (1937). Cited in Shamdasani, *Jung and the Making of Modern Psychology: The Dream of a Science,* 82.

90. In a famous article John Dewey called on scientists to separate investigations on "psychical events" and those on reaction responses: "To sum up: the distinction of sensation and movement as stimulus and response respectively is not a distinction which can be regarded as descriptive of anything which holds of psychical events or existences as such." Yet on many occasions Dewey's recommendations were ignored, and reaction time research spread beyond scientific settings. John Dewey, "The Reflex Arc Concept in Psychology," *Psychological Review* 3 (1896): 369.

91. For a history of psychotechniques see Anson Rabinbach, *The Human Motor: Energy, Fatigue and the Origins of Modernity* (Berkeley: University of California Press, 1990).

used "personal coefficient" tests "regularly conducted" in the "Physiological departments of our universities" to improve industrial efficiency. In his famous *Principles of Scientific Management* (1911) he recommended "laying off many of the most intelligent, hardest working, and most trustworthy girls merely because they did not possess the quality of quick perception followed by quick action."[92] Academic experimental psychologists turned to Taylor's work to explain how *they* could expand their work to industrial contexts.

New breeds of experimental psychologists dedicated to increasing industrial and military efficiency were inspired by Taylor's reaction time methods. They complemented chronometric techniques by enlisting other technologies, including cinematography.

In France, Jean-Maurice Lahy, a scientist and soldier, was one of the main advocates of Taylorism and one of the most important contributors to the growing field of psychotechnics. In 1908 Lahy was the director of the Laboratoire de psychologie expérimentale à l'École pratique des hautes études. Lahy, together with a group of young scientists, initially gathered in the laboratory of experimental psychology at the Asile de Villejuif to work with the experimental psychologist Édouard Toulouse.[93] They formed one of the most important reaction time research centers. Closely associated with the journal *Travail Humain,* these investigators studied reaction time to understand topics ranging from dactyloscopy to train conducting. They applied their research to professional selection and professional orientation.

At the outbreak of WWI, Lahy interrupted his studies on experimental psychology and Taylorism to study gunners fighting in the Argonne region. To "completely use our human resources," war needed to be "Taylorized."[94] Hoping to optimize the efficiency of battle forces in the same way that he had previously sought to increase the efficiency of industrial complexes, he studied their reaction time. His research was complemented by the work of the medical doctors Jean Camus and Henri Nepper, who also worked with reaction time during the war. In a set of pioneering studies they tested the reaction time of aviation pilots.[95] The numbers found by these investigators

92. Frederick Winslow Taylor, *Principles of Scientific Management* (New York: Harper and Brothers, 1911).
93. Édouard Toulouse headed the Laboratoire du service départemental de prophylaxie mentale.
94. Jean Maurice Lahy, "Sur la psycho-physiologie du soldat mitrailleur," *Comptes rendus des séances de l'Académie des sciences* 163 (1916): 33, and *Le Système Taylor et la physiologie du travail professionnel* (Paris: Masson, 1916), 69–70.
95. Jean Camus and Henri Nepper, "Mesure des réactions psychomotrices des candidats à l'aviation," *Paris médical* (18 March 1916), "Les Réactions psychomotrices et émotives des trépanés," *Paris médical* (3 June 1916), and "Temps de réactions psychomotrices des candidats à l'aviation," *Comptes rendus des séances de l'Académie des sciences* 163 (1916). During WWII, reaction tests on aviation pilots were incorporated into air force selection processes.

were much higher than those in Lahy's frontline experiments, but the extraordinary conditions under which the latter were made explained the discrepancy. Lahy's subjects fought for eighteen months in French lines that frequently came to within 50 meters of the German lines. A sympathetic reviewer of Lahy's work explained how "a minimum distraction could cost them their lives. These men had to evade a bullet as well as a trench shell or a few grenades. It is understandable that, under such conditions, these soldiers showed themselves ready to act with a singular vigor, in order to escape by their motor ability and the promptitude of their movements the perpetual swords of Damocles suspended over their heads!"[96]

At the war's end, Lahy returned to his original mission of improving industrial efficiency. He became the head of a new laboratory of psychotechniques in the Parisian transportation industry (Société des Transports en commun de la Région Parisienne, or S.T.C.R.P.) with more than thirty thousand employees under his direct supervision. Lahy built complicated machines to test the professional aptitudes of bus and tramway drivers by measuring their time of reaction.[97]

Lahy reacted against the use of the term "personal equation" for these studies. He noted how "W. Taylor gave the name of personal equation of an individual to that which we call reaction time" and asked, "Can we really call personal equation the duration of reaction time?"[98] He repudiated the term because it "presupposed that all the psychological characteristics of an individual intervene in its determination."[99] He separated the two terms, narrowly defining reaction time with military, industrial, and commercial concerns and associating the personal equation with psychology more broadly.

96. Octave Grimaud, "L'Examen psycho-physiologique des soldats mitrailleurs," *La Science et la vie* 13, no. 36 (1917–18): 125.

97. A driver under examination sat in the conductor's seat of a life-size bus replica. Three-hundred meters of film, designed to give the impression that "he was in a moving car," were projected unto a screen in front of him. The films used in these tests were specifically designed to "captivate their attention and move them emotionally." Others were composed of "cinematographic scenes [that] were changing, without a thread, such as the spectacles of the street seen through a car that moves and the memories that can present themselves to the spirit of a worker, even in the course of his shift." During the course of the spectacle, the candidate was asked to react to colored lights and sirens coinciding with certain cinematographic scenes. The time needed to react to these signals and film events was then recorded automatically. The best drivers were thus scientifically culled, and the proper functioning of modern transportation networks was assured. Jean Maurice Lahy, *La sélection psychophysiologique des travailleurs* (Paris: Dunod, 1927), 155, 72.

98. Taylor used the term "personal coefficient."

99. Lahy always insisted that reaction time was not the sole determinant of industrial efficiency. He criticized Münsterberg for believing that a "single test" could be used to analyze performance in a particular field and insisted on having a "plurality of tests." Lahy ignored that the critique of basing analysis solely on reaction time had already been leveled by Münsterberg, who proposed additional tests to measure attention. On Münsterberg's error, see ibid., 15. On a plurality of tests, p. 214.

Cultures of Reaction

The importance of reaction time spread from industry and the military to
other areas of modern life, characterizing its fast-paced task orientation.
In *The Will to Power,* a book attributed to Friedrich Nietzsche, the author
lamented how in modern times "men unlearn spontaneous action, they
merely *react to stimuli* from the outside."[100] Joseph Jastrow, author of the
Time-Relations of Mental Phenomena (1890), illustrated the pervasiveness
of modern, stimulus-response behavior:

> A great variety of actions may be viewed as responses to stimuli. There is a flash of
> light, and we wink; a burning cinder falls upon the hand, and we draw it away; a bell
> rings, and the engineer starts his train, or the servant opens the door, or we go down
> to dinner; the clock strikes, and we stop work, or go to meet an appointment. Again,
> in such an occupation as copying, every letter or word seen acts as a stimulus, to
> which the written letter or word is the response; in piano playing, and the guidance
> of complicated machinery, we see more elaborate instances of similar processes.
> The printer distributing "pi," the post-office clerk sorting the mails, are illustrations
> of quick forms of re-action, in which the different letters of the alphabet or the dif-
> ferent addresses of the mail matter act as stimuli, and the placing them in their ap-
> propriate places follow as a response. In many games, such as tennis or cricket, the
> various ways in which the ball is seen to come to the striker are the stimuli, for each
> variation of which there is a precise and complex form of response in the mode of
> returning the ball. In military drill the various words of command are the stimuli, and
> the actions thus induced the responses; and such illustrations could be multiplied
> indefinitely.[101]

The American experimental psychologist Edward Wheeler Scripture
described the widespread consequences of reaction time. To illustrate its
dangers, he imagined grouping the most important political figures of the
time (including Queen Victoria and Lenin) next to each other to form a
chain. If one of them perceived a stimulus, it would touch another one, who
would then touch another one, and so on. This model illustrated how reaction times of the order of a tenth of a second accumulated to trigger a "chain
reaction" with significant, perhaps disastrous, political consequences.

100. Friederich Nietzsche, *The Will to Power,* trans. Walter Kaufmann and R. J. Hollingdale (New York:
1967), 47. Italics mine. Cited in Jonathan Crary, *Suspensions of Perception: Attention, Spectacle, and Modern
Culture* (Cambridge, Mass.: MIT Press, 1999), 53. This aspect of modernity has been a source of criticism
for the New Left. See Max Horkheimer, "The End of Reason," in *The Essential Frankfurt School Reader,* ed.
Andrew Arato and Eike Gebhardt (New York: Continuum, 1982), 38.
101. Jastrow, *The Time-Relations of Mental Phenomena,* 4.

Figure 2.6. "Reaction chain" showing important political figures of the time: Lenin, Queen Victoria, etc. From Edward Wheeler Scripture, *Thinking, Feeling, Doing* (Meadville, Pa.: Chautauqua-Century Press, 1895), p. 39.

In 1934 the famous writer Aldous Huxley described a brave new world of reaction time in his travels in South America. Upon seeing two American ethnologists at work, he described the plight of their Guatemalan subjects: "Utterly miserable, but resigned, like sheep being led to the slaughter, half a dozen Indians permitted themselves to be taken, one by one, measured, weighed, tested for their reactions, inked for finger-prints. The spectacle was ludicrous and pathetic. It was absurd that people should make such an agonized face about so little. Absurd, and yet the agony was obviously genuine. These poor creatures really suffered from being just looked at."[102]

Instituting the "Standard Account"

Why, if the significance of reaction time and personal equation studies often spread beyond astronomy and experimental psychology, did most scientists see them largely in light of the 1796 incident between Maskelyne and Kinnebrook? The institutionalization of the "standard account" was due to Théodule Ribot, recipient of the first chair of experimental psychology and its main representative in France. Ribot's fame slowly eclipsed that of Charcot. In contrast to the notorious doctor from the Salpêtrière, Ribot defended a new type of psychological practice, which would eventually be

102. Aldous Huxley, *Beyond the Mexique Bay* (New York: Harpers and Brothers, 1934).

called experimental psychology and which was largely based on reaction time experiments. According to Ribot, reaction time studies could shed light on "l'ennui, le taedium vitae," which he considered to arise from a "languor of mental life."[103] They could be used to study "idiots," "cretins," "certain paralytics," and conditions where "the speed of thought" was considered to be remarkably slow.[104]

Ribot was essential in establishing reaction time experiments as post-Kantian. According to him, these new techniques helped rid psychology of the limitations imposed by harmful "kantian doctrines."[105] He actively criticized the "vagueness" of Kant's theories, "which left thought in a sort of mystical region where it seemed inaccessible to measurement, with respect to its duration."[106] The timed, temporal reactions to stimuli of an experimental subject were seen as opening a door toward increased quantification in the human sciences. A distinct breed of scientists saw themselves as finally prostrating the mysterious Kantian a priori. They saw themselves as rewriting the boundaries between physics and metaphysics and between the human and exact sciences.

Ribot added further pieces to the story of Maskelyne.[107] He proposed the claim, which has since then become a well-known trope, that the systematization of reaction time experiments was "for the most part" performed by German scientists and most systematically by Wundt. Ribot argued that French scientists had a slow awakening to the potential of reaction time studies. The fact that France had to wait until 1885 to obtain a new chair of experimental psychology at the Sorbonne and for the formation of the Société de psychologie physiologique has since been often cited as evidence of the discipline's underdevelopment.[108] To this day, the history of experimental psychology is usually traced to Germany, to Leipzig and to Wundt.[109]

---

103. Théodule Ribot, *La Psychologie allemande contemporaine (école expérimentale)* (Paris: Libraire Germer Baillière et Cie, 1879), 300.

104. Ibid.

105. Johann Friedrich Herbart (1776–1841), who filled the chair vacated by Kant in Königsberg, became the aim of similar criticisms. For criticisms of Kant and Herbart by Ribot see Robert H. Wozniak, "Théodule Armand Ribot: German Psychology of To-Day," in *Classics in Psychology, 1855–1914: Historical Essays* (London: Thoemmes Continuum, 1999; originally published, 1879; originally published in English, 1886).

106. For his criticisms of Kant, see Ribot, *La Psychologie allemande contemporaine (école expérimentale)*, 333, 64.

107. For Ribot on Maskelyne, see Théodule Ribot, "De la durée des actes psychiques d'après les travaux récents," *Revue philosophique de la France et de l'étranger, dirigée par Th. Ribot* 1 (1876): 268, and *La Psychologie allemande contemporaine (école expérimentale)*, 301.

108. Parot, "La Psychologie scientifique française et ses instruments au début du XXe siècle"; Danziger, *Constructing the Subject: Historical Origins of Psychological Research;* Brooks, "Philosophy and Psychology at the Sorbonne, 1885–1913"; Nicolas and Ferrand, "Pierre Janet au Collège de France."

109. Ash, "The Self-Presentation of a Discipline: History of Psychology in the United States between Pedagogy and Scholarship."

Although Ribot's neglect of French experiments was most apparent in his book on German psychology, he ignored work performed in France before and after this publication. Already in his first essay on German psychology he claimed: "The majority of works [on the physiology of sensations] have been done in Germany and are little known in France."[110]

Ribot interpreted Wundt in a manner that, according to his critics, left the philosopher-psychologist from Leipzig completely unrecognizable. His contemporaries even accused him of "inventing Wundt."[111] While in his early works Ribot included Wundt's work on aesthetics, morality, and religion, he increasingly focused on a smaller aspect of psychology and of Wundt's voluminous oeuvre. He slowly limited the vast field of German psychology to certain interpretations of the personal equation and to psychophysics, two different fields of knowledge that he increasingly classed together. Recent historians still note the strange historiographic "effect of removing from view a large part of Wundt's work, in particular his Völkerpsychologie."[112] Reaction time experiments were, in fact, never central to Wundt's own investigations. They merely constituted a preliminary and inferior entrance to the much broader field of ethics and folk psychology.[113]

A closer look at Ribot's oeuvre reveals that he ignored experimental work done in France—especially work based on alternative interpretations of

---

110. See Théodule Ribot, "La Psychologie physiologique en Allemagne," *Revue scientifique* 7 (1874): 553. In the next installment he tempered his earlier claim by writing: "Although [in Germany] there does not exist a proper school of psychology, and although they have not published complete and systematic treatises as the English, they have no less contributed a good number of new and truly scientific studies to psychology." Théodule Ribot, "La Psychologie allemande contemporaine: M. Wilhelm Wundt," *Revue scientifique* 8 (1875): 723. In his later work he also included the caveat that research on the duration of mental acts was "not the exclusive concern of German physiologists." Ribot, *La Psychologie allemande contemporaine (école expérimentale)*, 299. But later he would continue to proclaim Wundt and Germany as advocates of experimental psychology, and France as against it. See Théodule Ribot, "Histoire des sciences: Leçon d'ouverture du cours de psychologie expérimentale et comparée du Collège de France," *Revue scientifique* 15 (1888), and "Psychologie: La Psychologie physiologique en 1889," *Revue scientifique* 18 (1889). Other articles included Ribot, "La Psychologie allemande contemporaine: M. Wilhelm Wundt," "La Psychologie physiologique en Allemagne: M. W. Wundt," *Revue scientifique* 9 (1875), and "La psychologie physiologique en Allemagne: M. W. Wundt." In his university course on "les états inconscients," Ribot lectured on the concepts of fusion and simultaneity of sensation but again did not mention French authors, stressing instead the work of Helmholtz and Wundt. Ribot, cours, Ms. 2354, Archives de la Sorbonne, pp. 10–11.
111. Fr. Picavet, "Philosophes français contemporains, M. Théodule Ribot," *Revue politique et littéraire* 2, no. 18 (1894): 592.
112. Carroy and Plas, "The Origins of French Experimental Psychology: Experiment and Experimentalism," 77.
113. For one of the first attempts to question the association of experimental psychology and Wundt, see Arthur L. Blumenthal, "A Reappraisal of Wilhelm Wundt," *American Psychologist* 30 (1975). Also see Martin Kusch, "Recluse, Interlocutor, Interrogator: Natural and Social Order in Turn-of-the-Century Psychological Research Schools," *Isis* 86 (1995). While Ribot tried to bring German work to the French public, his German sources ironically reached him through a frenchified lens. For example, for his account on Fechner and psychophysics he followed their interpretation by Joseph Delboeuf, the Belgian psychophysicist. Delboeuf's work appeared to Ribot "clearer than Fechner's." Ribot, "La Psychologie physiologique en Allemagne," 555 n. 1.

reaction time and the personal equation. Readers criticized his focus on Germany and could not understand why he did not trace the history of experimental psychology to France. The total absence of French experimental psychology in Ribot's work astounded critics. A review that appeared in the famous *Revue philosophique* (which Ribot founded and edited) could not help but complain how "Ribot tends to confuse metaphysical psychology with French psychology."[114] His focus on Germany was clearly partisan. Since Ribot had published one book on English psychology and one on German, the bewildered reviewer asked, "Why doesn't he honor the French in the same way?"[115] While some complained of Ribot's bias, others, like Pierre Janet, who would hold chairs of experimental psychology at the Sorbonne and at the Collège de France after Ribot, lauded his strategy and called it a "testimony, in any case, of a certain perspicacity."[116]

Yet despite the prevalence of reaction time experiments in many areas of science, a frequently noted characteristic of institutional French experimental psychology was its paradoxical dearth of experimentation. Ribot held a degree in philosophy and was not trained in laboratory techniques. Although he performed almost no experiments, he—ironically—came to symbolize French experimental psychology. While disdaining philosophers as "people for whom the best part of the cake is that which they cannot eat,"[117] he did not include any laboratory practices in his course. Only when he moved to the Collège de France did he finally ask for a laboratory.[118] Unqualified for doing experiments himself, he named Henri Beaunis its director in 1889. Beaunis, however, also abandoned the laboratory a few years after his appointment and left the spoils for Alfred Binet, his assistant. In career terms, the laboratory that had already been abandoned by Ribot and Beaunis was a sinking boat. The experimentally skilled Binet eventually lost against the philosophically inclined Pierre Janet when the bid for the chair reopened after Ribot retired.

While Ribot's account was canonized, others were forgotten. Indeed, forgetting was essential for establishing experimental psychology as a distinct discipline. Other writers, including Wundt himself, found origins for

---

114. Thomas Victor Charpentier, "Th. Ribot.—La Psychologie allemande contemporaine," *Revue philosophique de la France et de l'étranger, dirigée par Th. Ribot* 9 (1880): 351.

115. Ibid., 355.

116. Pierre Janet, "La Psychologie expérimentale et comparée," in *Le Collège de France (1530–1930), Livre jubilaire composé à l'occasion de son quatrième centenaire* (Paris: Presses Universitaires de France, 1930), 226.

117. Ribot, 15 September 1879, Bibliothèque Victor Cousin, Correspondance de Lionel Dauriac, Vol. 4, M-R, letter no. 144-1045, p. 633, on pp. 2–3.

118. John I. Brooks III, "Philosophy and Psychology at the Sorbonne, 1885–1913," *Journal of the History of the Behavioral Sciences* 29 (April 1993).

experimental psychology that differed markedly from the ones described in the "standard account." Its origin was not always traced back to 1796, to astronomy, or to Wundt. Even Wundt reacted against the association of experimental psychology with himself and Germany, for which he blamed Ribot. Although he called Ribot's *La Psychologie allemande* a "very remarkable and clear exposition," he had one objection: "But, on only one point, the description of M. Ribot can, I believe, give ground to a wrong interpretation." Ribot exaggerated the experimental side of German psychology by considering it the "general or the only preponderant" method and by giving it an "excessively prominent" place in his book. Experimental psychology, according to Wundt, was in fact "detested" in Germany, even considered "a blasphemy."[119] But historians of psychology continued to consider reaction time experiments as foundational for German experimental psychology.

Many of Ribot's assertions were highly polemical. Did reaction time experiments herald a post-Kantian era where quantification was extended to include even the psyche? Did the personal equation reveal individual traits, such as personality, nervous constitution, age, health, educational level, mental makeup, and intelligence? How were reaction time and the personal equation related to a person's appearance, physiognomy, and race? Who and where were the French astronomers and physicists ignored by Ribot?[120] Experiments on reaction time and the personal equation were hardly absent in France.[121] The philosopher and musician Lionel Dauriac tried to include them. He reminded readers how physicists always measured the effect of transmission delays on their results, including those of reaction time: "The experiments [described by Ribot] . . . are of the same nature as physics experiments: when one experiments at Montlhéry and at Villejuif on the speed of sound, one proceeds, although on a vaster scale, as proceeds Wundt and Donders. Experimenters . . . always account for the duration

---

119. Wilhelm Wundt, "Préface de l'auteur pour l'édition française," in *Éléments de psychologie physiologique* (Paris: Félix Alcan, 1886).

120. When Ribot included the work of French experimenters he often ignored their own conclusions and presented their findings in manner consonant with his own interpretation. For example, he cited the astronomer Charles Wolf's claims that the personal equation could be reduced through habit and attention, but completely ignored his weightier claim that the remaining personal equation was entirely retinal. Ribot's elision of French work contrasted sharply with the influential account of the personal equation by the American psychologist Edmund C. Sanford, editor of the *American Journal of Psychology*, which focused largely on France and on the work of Wolf.

121. This experimental tradition notwithstanding, Ribot and Beaunis repeatedly asserted that experimental psychology was mainly a Germanic discipline. Beaunis kept competing French experiments on sensation out of his influential pedagogical treatises until 1888, when he dedicated to them an unflattering footnote. His veiling of early experiments performed in France was clearly successful. At the time of his death he was remembered as "the first in France to pave the way for physiological psychology." "Chronique," *L'Année psychologique* 22 (1920–21).

of nervous transmission as a forced physiological intermediary."[122] In addition to the experiments on acoustics described by Dauriac, optics, ballistics, astronomy, photometry, metrology, and measurements on the speed of light, all dealt with physiological intermediaries. The characterization of the personal equation and reaction time as measurements related to the speed of thought was frequently contested. Ribot acknowledged some of the difficulties (such as how results varied among different experimenters)—but not all. Provocatively, he claimed these "hardly mattered, since the essential has been established: the possibility of measurement."[123]

Ribot's account of precision measurement spreading from the physical sciences to the human sciences was frequently called into question. His optimistic portrayal of a revolutionary moment when even the most recalcitrant object, thought, had been conquered, contrasted starkly with the work on this topic by other scientists, characterized by debates and difficulties in the establishment of the most basic time and length measurements. These difficulties affected mundane scientific practices, such as mapmaking and navigation, the establishment of time, and the determination of the length of the meter. Additionally, they cast in doubt some of the most important constants informing scientists' knowledge of the universe.

Through reaction time and the personal equation, widespread attention to tenth-of-a-second moments spread far from astronomy, far from experimental psychology, and far from the "standard account" of their history.

122. Lionel Dauriac, "Le Mouvement philosophique: De la psychologie expérimentale en Allemagne," *Revue politique et litteraire* 17 (1879): 242.
123. Ribot, "De la durée des actes psychiques d'après les travaux récents." 287.

CHAPTER 3

THE MEASURE OF ALL THOUGHTS

An elementary cerebral vibration has a certain duration, and this duration is approximately a *tenth of a second.*

CHARLES RICHET, Nobel Prize winner and physiologist

During the second half of the nineteenth century, influential physiologists and psychologists came to the conclusion that the tenth of a second was a constitutive unit of human consciousness. One of the most important scientists who stressed the significance of this moment was Charles Richet, founder of the Société de psychologie physiologique and (later) one of the most famous scientists of France, earning the Nobel Prize in Medicine in 1913.[1] Richet started his career by studying reaction time and publishing novels under the pseudonym Charles Epheyre. By the turn of the century, he was a well-established figure and author of one of the most important textbooks of physiology. In it, under the entry for "brain," he stressed the importance of the tenth of a second:

1. Jean-Martin Charcot was president of the Society (although not actively involved). Pierre Janet and Théodule Ribot were vice-presidents, and Richet was secretary. Richet also founded the *Annales des sciences psychiques,* which later became the *Revue métapsychique.*

> It seems we have the right to consider the *tenth of a second* as the psychological unit
> of time for conscious phenomena, from the point of view of volition or perception. . . .
> In other words, an elementary cerebral vibration has a certain duration, and this
> duration is approximately a *tenth of a second*. A discontinuous cerebral fact cannot be
> distinguished if the intervals that separate the elementary reactions are less than a
> *tenth of a second*. If they are smaller, discontinuous facts become continuous.[2]

According to Richet, even the smallest combustive explosions lasted about as long as the shortest time of reaction: both were approximately a tenth of a second.[3]

How did this unit of time become associated with mental, physical, and even chemical processes? To answer this question, we must delve into the late nineteenth-century history of psychology and physiology and focus on an important yet previously unstudied debate surrounding reaction time experiments. The stakes of this debate were high, involving essential revisions to the concept of experimentation and to the place of the experimenter in it.

The increase of quantification's reach into tenth-of-a-second mental processes was highly contested. At the center of these contestations stood an important question: Was an experimental system composed of stimulus, experimental subject, and recording device valid? The two principal scientists working on reaction time in France reached opposite conclusions. One of these scientists was Richet, who defended the view that reaction time measured the speed of thought. He first measured the "personal equation" in persons suffering from ataxia, but his concerns were not only medical.[4] He called on experimental psychologists to follow the criminologist Cesare Lombroso's attempt at "finding the limits which distinguish the insane from the criminal."[5] For him studying human factors *first* was so important that all other investigations, including astronomical or cosmic ones, were subordinate: "Psychology . . . that is humanity in its entirety . . . is the only true KOSMOS. Why should we care about the moon and the stars?"[6] For A.-M. Bloch, a lesser-known physiologist who would become vice-president

---

2. Charles Richet, "Cerveau," in *Dictionnaire de physiologie*, ed. Charles Richet (Paris: Alcan, 1898), 10–11. Italics mine.

3. Richet argued that a cerebral, tenth-of-a-second process "is without doubt a phenomenon analogous to an explosion" that "needs a certain time to occur." Ibid., 17.

4. Charles Richet, "Études sur la vitesse et les modifications de la sensibilité chez les ataxiques: Note lue à la Société de biologie, dans sa séance du 17 juin 1876," *Comptes rendus des séances et mémoires de la Société de biologie* 3 (1877). He experimented on patients from the Salpêtrière and the Hôtel-Dieu. For these measurements he used the graphic method learned from Marey and was assisted by Jules Luys.

5. Charles Richet, "L'Avenir de la psychologie," *Revue scientifique* 50 (3 September 1892): 293.

6. Ibid., 295.

of the Société de biologie, reaction time proved the essential unquantifiability of voluntary action.[7] His results led him to reevaluate basic assumptions about time and their relation to the present, concluding, in astonishment, that a "mathematically actual phenomenon" simply "did not exist."[8] Richet's unwavering belief in a thoroughgoing materialism and determinism remained irreconcilable with Bloch's belief in the indeterminacy of voluntary action.

The validity of reaction time experiments was frequently debated in these years. Shortly after Bloch published his results, the engineer Léon Lalanne joined the debate, explaining the results of experiments performed thirty-four years earlier. These started with the "vulgar and probably known since antiquity . . . children's game" consisting of whirling a glowing ember from a string and seeing it form a closed circle: "The luminous circle is completely closed once the speed of rotation reaches ten [complete circles] per second."[9] Does a "tenth of a second elapse from the moment [the ember] leaves one of the points to the moment it comes back to it?"[10] He then extended these investigations to tactile persistence, concluding that sensations appeared continuous if these succeeded each other at intervals that "never surpassed 1/10th of a second."[11] Extrapolating to sound, he determined that "certain uniform and isochronic noises fuse into a unique sound and one is authorized to say, that as in the experiments of the luminous circle, the persistence of sensation lays approximately between 1/8th to 1/10th of a second."[12] As the debate about the significance of these experiments raged, investigators nonetheless became more and more aware of the significance of this value.

7. Adolphe-Moïse Bloch was a doctor at the Asile de Vincennes, member of the Société de biologie, and laureate of the Institut de France. In 1903 he became vice-president of the Société de biologie. The motor used for his 1875 experiments was installed in the laboratory of physiology of the Muséum d'histoire naturelle. He initially worked in Claude Bernard's laboratory at the Collège de France. Bloch's accomplishments included building a widely used *sphygmomètre*. See Charles Féré, "Observations faites sur les épileptiques à l'aide du sphygmomètre de M. Bloch," *Comptes rendus des séances et mémoires de la Société de biologie* 5 (1888).

8. A.-M. Bloch, "Psychologie: La Vitesse comparative des sensations," *Revue scientifique* 39 (7 May 1887): 585.

9. Léon Lalanne, "Note sur la durée de la sensation tactile," *Journal de l'anatomie et de la physiologie normales et pathologiques de l'homme et des animaux* 12 (1876): 453. A shorter version of this article appeared in Léon Lalanne, "Sur la durée de la sensation tactile," *Comptes rendus des séances de l'Académie des sciences* 90 (1876). It was reviewed in "Journal de l'anatomie et de la physiologie, etc., L. Lalanne. Note sur la durée de la sensation tactile," *Revue philosophique de la France et de l'étranger, dirigée par Th. Ribot* 1 (1876). Lalanne complained how no distinction was made between "the speed at which an exterior impression reaches the brain" and "the duration of perception." He insisted that these two phenomena were "completely different." Lalanne, "Note sur la durée de la sensation tactile," 450.

10. Lalanne, "Note sur la durée de la sensation tactile," 453.

11. Ibid.

12. Ibid.

The physiologist Étienne-Jules Marey tried to solve the impasse between these two positions. He reviewed the work of Richet and Bloch and concluded that, despite their differences, they both agreed about the existence of a tenth-of-a-second delay between stimulus and response: "One sees that the signal is delayed on average a 1/10 of a second from the moment of stimulation; that is the measurement of a personal error."[13] Marey was famous for developing the graphic method in science that had been pioneered by the physiologist Carl Ludwig, by Hermann von Helmholtz, and others. Since 1851 Helmholtz had argued that "if intervals of less than *one-tenth of a second* are to be observed with accuracy, or even measured, we must have recourse to artificial means,"[14] and Marey followed his recommendation by further developing graphic methods that he described as "superior to all other forms of expression" and revealing the "language of the phenomena themselves."[15]

The graphic method, however, did not eliminate the debate surrounding these experiments and in fact caused additional problems. Other techniques, such as photographs and direct reading instruments, were soon enlisted. Yet these did not solve the controversy either.

In analyzing controversies in science, sociologists of science have noticed that "experiments in real, deeply disputed, science hardly ever produce a clear-cut conclusion."[16] The debate between Richet and Bloch confirms this assessment. Yet it also reveals an additional lesson. By the time their debate ended, these scientists transformed the concept of experimentation itself.[17] They installed the stimulus-subject-clock circuit as a viable experimental system that would characterize an increasingly cybernetic world.

During the second half of the nineteenth century, physiology and psychology changed to an unprecedented degree. A new focus on experiments—particularly on reaction time—gained ground over the clinical psychology associated with Jean-Martin Charcot. New laboratories were built and new techniques were developed across Europe and North America based on new relations between psychologists, experimental subjects, and instruments.

13. Étienne-Jules Marey, *La Méthode graphique dans les sciences expérimentales et principalement en physiologie et en médecine* (Paris: G. Masson, 1878), 151.

14. Hermann von Helmholtz, "Ueber die Methoden, kleinste Zeittheile zu messen, und ihre Anwendung für physiologische Zwecke," *Königsberger Naturwissenschaftliche Unterhaltungen* 2 (1851): 325, and "On the Methods of Measuring Very Small Portions of Time, and Their Application to Physiological Purposes," *The London, Edinburgh and Dublin Philosophical Magazine and Journal of Science* 4 (1853): 189. Italics mine.

15. Marey, *La Méthode graphique dans les sciences expérimentales et principalement en physiologie et en médecine*, iii.

16. Harry Collins and Trevor Pinch, *The Golem: What You Should Know about Science* (Cambridge: Cambridge University Press, 1998), 147.

17. For an analysis of the changing notion of experimentation, see Peter Galison, *Image and Logic: A Material Culture of Microphysics* (Chicago: University of Chicago Press, 1997).

Some scientists, such as Franz Brentano, the famous precursor to phenom-
enology, reacted against the growing focus on reaction time experiments
for studying the mind and lamented the concomitant neglect of introspection techniques. He described how enemies of introspective psychology
buttressed their case with the example of the personal equation: "They
point to the remarkable case of the error that occurs in the determination
of time. This error regularly appears in astronomical observations in which
the simultaneous swing of the pendulum does not enter into consciousness simultaneously with, but earlier or later than, the moment when the
observed star touches the hairline of a telescope."[18] Yet most defended the
new discipline and its focus on reaction time experiments.

Since the 1860s the clinical investigator Duchenne de Boulogne showed
how electric currents applied to a person's skin caused underlying muscles
to expand or contract. His experiments prompted a commentator to re-
joice in how electricity had just "replaced the scalpel."[19] As the century progressed, the ambitions of physiologists intensified along with an increasing
use of electric technologies. They extended their experiments by testing the
effects of electricity not only on motor nerves but also on sensory nerves.
They incorporated a subject's reaction. They deepened the reach of the electric scalpel by measuring the speed of mental processes.

Reaction Chain

Bloch found a contradiction in the numbers for the speed of sensory transmission derived from two competing methods.[20] Reaction time experiment
gave a result of around 65 meters per second. The other method based on
the phenomenon of tactile persistence, resulted in 194 meters per second.

18. Franz Brentano, *Psychologie du point de vue empirique*, ed. L. Lavelle and R. Le Senne (Paris: Aubier,
1944), 108. Original published in 1874. Wilhelm Wundt used the personal equation as evidence to prove
that the mind could not focus on two things at the same time—for example, on consciousness and on
how consciousness worked. This was used to prove that certain introspection techniques were bound
to fail. Wundt eventually abandoned this view, to which he retrospectively (and disdainfully) referred as
the "needle's eye theory of consciousness." See Solomon Diamond, "Wilhelm Wundt," in *Dictionary of
Scientific Biography* (New York: 1970), 527. Ernst Mach criticized this aspect of Wilhelm Wundt's work by
reference to the personal equation; see Claude Debru, "Ernst Mach et la psychophysiologie du temps,"
Philosophia Scientiae 7, no. 2 (2003): 72.

19. Amédée Latour, "Revue générale: Mécanisme de la physionomie humaine," *L'Union médicale* 103 (2
September 1862): 417. Cited in Alisa Luxenberg, " 'The art of correctly painting the expressive lines of the
human face': Duchenne de Boulogne's Photographs of Human Expression and the École des Beaux-Arts,"
History of Photography 25 (2001): 203.

20. A.-M. Bloch, "Expériences sur la vitesse du courant nerveux sensitif de l'homme," *Archives de physiologie normale et pathologique, publiées par MM. Brown-Séquard, Charcot, Vulpian* 2 (1875). A short paragraph
mentioning Bloch's work appeared in "Archives de physiologie normale et pathologique, M.A. Bloch.
Expériences sur la vitesse du courant nerveux sensitif de l'homme," *Revue philosophique de la France et de
l'étranger, dirigée par Th. Ribot* 1 (1876).

The contradiction in the two numbers existed, Bloch explained, because scientists could not agree about how to separate sensory, motor, and mental phenomena.[21]

The physiologist disagreed with the common interpretation of reaction time experiments as measurements of the speed of thought. Working from the Muséum d'histoire naturelle, he hoped to demonstrate "the inanity of the experiments conceived from this type."[22] In order to bypass the variations arising from cerebral processes, Bloch measured the speed of sensory transmission without having the subject react. By avoiding the time of reaction he believed he could "avoid the operation of the intellect" and study "pure sensibility."[23] When Bloch applied approximately fifty successive shocks per second to the skin, the sensation of discreteness disappeared. The successive shocks merged into a feeling of continuous impression. The time between impressions was much shorter than that furnished by reaction time experiments. These results were paradoxical. How could the time taken by impressions to reach the sensorium in experiments on sensory persistence be so different from the one derived by reaction experiments? Which of these two methods should be trusted? The first one, according to Bloch, since it did not require the subject to react. It, therefore, did not introduce unpredictable elements into the experiment.

Bloch's views on reaction were consistent with the philosophy of Victor Cousin, which placed strict limits on scientific materialism by dividing the human body into passive and active components. It identified nature with vegetarian passivity, sensibility, and femininity. Selfhood was in turn identified with activity, willfulness, and masculinity.[24]

---

21. Bloch considered that the only senses relevant to these types of investigations were the "passive" ones of hearing, touch, and sight, since smell and taste required "volitional" acts of breathing and swallowing.

22. Bloch, "Expériences sur la vitesse du courant nerveux sensitif de l'homme," 592.

23. Ibid., 599. Bloch applied a light shock to a part of the body (for example, the right index finger) and then quickly applied another shock to a homologous part of the body (for example, the left index finger). He noticed that if these shocks were in fast succession, the subject would sense them as simultaneous. Then he measured the time at which successive shocks applied to different body parts, like the ear and the finger, were perceived as simultaneous by the subject, and, by subtracting this from the previous number for two fingers, he came up with a radically different number for the speed of sensory transmission. He distinguished the different times it took for an impression to travel through the nerve cords or through the spinal cord. By experimenting on body parts with different sensibilities, like the palm of the hand and a finger, he made sure his results were not affected by the difference in the time in which an impression was received by the nerves, at the level of the skin. With full confidence, Bloch gave a new number for the speed of sensory transmission. He concluded that the speed of sensory transmission was 194 meters per second for the spinal cord and 134 meters per second for the nerves. His average result was 156 meter per second.

24. For Cousin, the self and the will were identical, mutually defining each other: "Our personality is the will, and nothing more." Cousin, "Préface à la première édition," *Fragmens philosophiques*, 2nd ed. (Paris: Ladrange, 1833), 39. Cited in Jan Goldstein, "Mutations of the Self in Old Regime and Postrevolutionary

Bloch's objections brought reaction time experiments to the limelight, since they showed potential errors in preexisting calculations for the speed of thought. The physiologist undertook further reaction time experiments on himself, where he encountered more paradoxes.[25] He noticed that he reacted more quickly to a stimulus applied to the hand than the nose, although the nose was certainly closer to the sensorium. The reason why subjects reacted more quickly to an impression on the face than on the foot could not be due to "the difference in nerve lengths" as the experts on the topics (Helmholtz, Emil du Bois-Reymond, Johan Jacob de Jaager, Franciscus Cornelis Donders, Rodolphe Radau, and Marey) had all believed. Other explanations had to be adduced. Times of reaction varied greatly according to his posture, his momentary disposition, the "progressive education" undergone by him on a daily basis and even in a single experiment, and if the part of the body under stimulation was more or less habituated to tactile impressions. These strange variations rendered reaction time experiments "unfit for rigorous research."[26] The reason why scientists could not agree on a precise figure for the speed of sensory transmission using reaction experiments was that "the real cause of these discordant results lies in the inherent variability of phenomena of an intellectual order."[27]

These denunciations were consonant with earlier criticisms of reaction time experiments. In 1859 the famous medical doctor Charles-Edouard Brown-Séquard, known for his work on endocrinology and on the intercrossing of sensory fibers in the spinal cord, attempted, unsuccessfully, to introduce them into physiology.[28] These chronometric investigations were part of his classic research on nervous transmission developed mostly through vivisection and through observations of clinical, pathological cases, but in these cases he worked without a scalpel.[29] Brown-Séquard started one chronometer along with an electric shock and had a person stop another chronometer after feeling the shock. The difference in the time of the two chronometers would then consist of five elements, analogous to the ones described by Helmholtz: "1° the time of centripetal transmission; 2°

France: From Ame to Moi and to Le Moi," in *Biographies of Scientific Objects*, ed. Lorraine Daston (Chicago: University of Chicago Press, 2000), 105.

25. Bloch used the light shocks of a thin feather and a water motor connected to the reservoir that served the Jardin des plantes. He performed these experiments on himself.

26. Bloch, "Expériences sur la vitesse du courant nerveux sensitif de l'homme," 599.

27. Ibid.

28. "Physiologie du système nerveux," *Journal du progrès des sciences médicales et de l'hydrothérapie rationnelle* 4 (1859): 323. Charles-Éduard Brown-Séquard used two chronometers with electromagnets.

29. In contrast to other contemporary experimenters, including Helmholtz, Brown-Séquard distinguished between the different speeds of transmission in the nerves and in the spinal cord and criticized previous work for not taking this difference into account.

the time of perception; 3° the time of the reflection produced by the will to stop the second chronometer; 4° the centrifugal transmission of this willed action through the muscles; 5° the time of muscular contraction."[30] Like Helmholtz, Brown-Séquard used the method of differential nerve lengths and explained how the speed of sensory transmission could be separated from the other elements by experimenting with different body parts.

Brown-Séquard's experiments attested to the growing realization that reactions to stimuli suffered from a lag time between the moment a phenomena actually occurred and the time it was perceived. Whenever a person tried to stop a timekeeping device at a given signal, this lag time surfaced. But reaction time experiments encountered a formidable resistance from members of the Société de biologie. His experiments did not convince "many" members of the Société de biologie because they noticed "enormous" differences between the numbers obtained from different experimental subjects. Brown-Séquard sought to "remedy" the problem by arranging people in a human chain instead of experimenting only on single individuals. The first person in the chain was asked to react to an electric shock at the same time that a chronometer would be set in action. Then this person touched a second one, who, in turn touched a third, who would then stop the chronometer. Theoretically, any number of individuals could be used, and a number for the speed of sensory transmission could be obtained by dividing the final result by the number of individuals in the human chain.[31] This solution did not satisfy his public, and after these isolated attempts Brown-Séquard abandoned work in chronometry and returned to what he did best, vivisecting animals with sartorial skill.

A Dangerous Materialism

Decades later, critics continued to question both the validity of these experiments and their interpretation as measurements of the speed of thought—an interpretation that many associated with a dangerous materialism.[32]

30. "Physiologie du système nerveux," 323. A summary of Brown-Séquard's work on sensation can be found in Brown-Séquard, "Recherches sur la transmission des impressions de tact, de chatouillement, de douleur, de température et de contraction (sens musculaire)," *Journal de la physiologie de l'homme et des animaux, publié sous la direction du docteur Brown-Séquard* 6 (1863).

31. Brown-Séquard's attempt to eliminate individual differences through the inclusion of more individuals in the reaction circuit differed from a statistical approach, by which the results obtained from different isolated individuals would be averaged.

32. The antagonism of certain conservative members of the universities to the growing materialism in the Paris Faculty of Medicine came to light during the *affaire Grenier*, in which the reformist Minister of Public Instruction Victor Duruy was pressured to annul the certification of a medical thesis on free will. This antagonism started to peter out by the late 1870s, when the Republic was "conquered by the republicans,"

A writer for *Les Mondes*, a conservative journal edited by a Jesuit priest, criticized the reaction time experiments of Donders, insisting that the soul was responsible for the perception of sensations, and it was certainly not located exclusively in the brain: "It is the soul's union with the body, and not the body alone which perceives sensations. And, the soul is not more in the head than in the feet; it is united to all the body and is present entirely in each part of the body."[33] Another article in *Les Mondes* noted the "great error" in calling the measured phenomena "mental," since Donders was really measuring nothing more than "simply a physical-physiological act."[34] Perceptions, critics argued, were not transmitted to the brain in a telegraphic manner.

Ramon de la Sagra, a conservative political commentator and scientist, ridiculed Donders's results. The partisan interpretation of reaction time experiments as mental "had its source in the essential materialism of his theory."[35] Donders did not measure "the duration of perceptions, of thought, or volition, which are truly intellectual actions." He was rather merely measuring "nerve movements, nerve transmissions, in a word: material phenomena." De la Sagra argued that intellectual phenomena "resist every material means of measurement, determination, or registration."[36] A reviewer of Donders's work objected to the word "choice" used to describe experiments in which an observer was asked to react to certain impressions with the right hand, and to others with the left. The time of each hand's reactions might not reflect the time of choice, but rather that a better exercised arm could react more quickly than a less exercised one.[37] Donders, another

---

Jules Ferry and Léon Gambetta. Jan Goldstein, "The Hysteria Diagnosis and the Politics of Anticlericalism in Late Nineteenth-Century France," *Journal of Modern History* 54 (1982): 226–27.

33. "It is thus evident that all of the investigations of this type amount to nothing else but conjectures." R. P. A. Bellynck, "Correspondance des Mondes: R. P. A. Bellynck, à Namur," *Les Mondes: Revue hebdomadaire des sciences et de leurs applications aux arts et à l'industrie par M. l'abbé Moigno* 19 (1869): 558.

34. "Accusés de réception: Archives néerlandaises des sciences exactes et naturelles," *Les Mondes: Revue hebdomadaire des sciences et de leurs applications aux arts et à l'industrie par M. l'abbé Moigno* 20 (1869). The journal *Les Mondes* was edited by the Abbé Moigno. For his role as editor and science journalist, see Maurice Crosland, "Popular Science and the Arts: Challenges to Cultural Authority in France during the Second Empire," *British Journal for the History of Science* 34 (2001). Articles that appeared in *Les Mondes* stood in sharp contrast to the places where Marey and Radau published: *Le Moniteur scientifique*, the *Revue des deux mondes*, the *Bulletin hebdomadaire de l'Association scientifique de France*, and the *Revue scientifique*. Charles Richet was a major shareholder of the *Revue des deux mondes* and participated in directing its content (his sister was married to its editor, Charles Buloz). He became editor of the *Revue scientifique* in 1881. Stewart Wolf, *Brain, Mind, and Medicine: Charles Richet and the Origins of Physiological Psychology* (New Brunswick, N.J.: Transaction Publishers, 1993), 56, 90.

35. Ramon de la Sagra, "Académie des sciences: Séance du lundi 1er février," *Les Mondes: Revue hebdomadaire des sciences et de leurs applications aux arts et à l'industrie par M. l'abbé Moigno* 19 (1869): 213.

36. Ibid., 214.

37. Le Cyre, "Dépouillement des journaux étrangers, par nos traducteurs volontaires: Durée de la transmission des sensations," *Les Mondes: Revue hebdomadaire des sciences et de leurs applications aux arts et à l'industrie par M. l'abbé Moigno* 19 (1869).

critic complained, ignored individual differences in the sense of touch. The tactile sense of people accustomed to "hard work," for example, was blunt compared to that of others. Yet another writer criticized Donders's bold use of the prefix "noë" (which he defined as soul in Greek) in the names of his two instruments: noëmachrograph and noëmachrometer.

Eminent scientists continued to ignore the contradictions revealed by critics. Ribot, who carried the flag of experimental psychology in France, ignored Bloch's conclusions.[38] He continued to claim reaction time experiments could be used to measure the speed of thought. In the very first volume of a new journal edited by him, the famous *Revue philosophique,* he published a provocative essay on reaction time titled "De la durée des actes psychiques" (1876).[39]

Richet similarly dismissed Bloch's results. Without acknowledging any controversy surrounding the value of the speed of sensory transmission, he took it to be 30 meters per second, claiming that "from the positive results of physiology, is roughly exact."[40] In his thesis on sensation, he adjusted the speed of sensory transmission to 50 meters per second, but continued to ignored Bloch's work because his "number is so different than the rest." He confidently stated that Bloch's criticisms "are probably wrong."[41]

That same year, Marey invited Richet and Bloch, the two experimenters who represented opposing positions, to study the question in his laboratory. The matter was urgent, since their experimental results differed greatly. For Marey, the problem was especially pertinent because, since very early on, he had associated his own graphic work with that of Donders and the astronomers. Trying to reach a common ground, Bloch and Richet adapted their methods.[42] After working together, Richet realized the difficulties in

38. Bloch's work was more favorably received in Germany than in France. Johannes von Kries, professor of physiology at Freiburg, and his collaborator Felix Auerbach stood by Bloch's "theory" of "how very uncertain the reaction method is for determining the speed of nervous transmission." These experimenters, who worked in Helmholtz's laboratory, explained how this "was found earlier by Bloch, and before him by Exner, that the reaction-time of a stimulus applied to the nose (the forehead in Exner) is longer than when the hand is stimulated, despite of the shorter route of transmission." Johannes von Kries and Felix Auerbach, "Die Zeitdauer einfachster psychischer Vorgänge," *Archiv für Physiologie: Physiologische Abtheilung des Archives für Anatomie und Physiologie, herausgegeben von Dr. Emil du Bois-Reymond* (1877): 357.

39. Théodule Ribot, "De la durée des actes psychiques d'après les travaux récents," *Revue philosophique de la France et de l'étranger, dirigée par Th. Ribot* 1 (1876). A reference to Bloch's work appeared in a footnote. Bloch's criticisms to the method of differential nerve lengths were ignored, as were his claim that including the element of reaction in the experiment voided any measurements.

40. Richet, "Études sur la vitesse et les modifications de la sensibilité chez les ataxiques: Note lue à la Société de biologie, dans sa séance du 17 juin 1876," 83.

41. Charles Richet, *Recherches expérimentales et cliniques sur la sensibilité* (Paris: G. Masson, 1877), 51.

42. To make his experiments more comparable to others, Bloch first changed his mechanical shocks to electric ones. In doing so, he noticed a change in his results. While in his previous method simultaneity depended on the tactile sensibility of the body part under study, electric shocks followed no such order. Also, he found different numbers for the speed of sensory transmission when he tested the feeling of

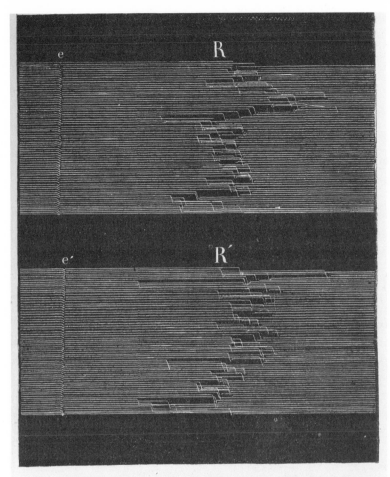

Fig. 69. Mesure de l'erreur personnelle, par Bloch. (Série supérieure *e*, instants où l'on a excité l'épaule ; R, instants où l'on a réagi à l'excitation. Série inférieure *e'*, excitation de la main ; R', réaction.)

Figure 3.1. Graph showing variations in measurements of personal errors. From E. J. Marey, *La Méthode graphique dans les sciences expérimentales et principalement en physiologie et en médecine* (Paris: G. Masson, 1878), p. 151.

fusion that appeared when he experimented on the same body part or the feeling of simultaneity when experimenting on different body parts. Some of the results of this collaboration were discussed in A.-M. Bloch, "Expériences sur la vitesse relative des transmissions visuelles, auditives, et tactiles," *Journal de l'anatomie et de la physiologie normales et pathologiques de l'homme et des animaux* 20 (1884), and "Caractères différentiels des sensations électriques et tactiles," *Travaux du laboratoire de M. Marey* 3 (1877). Experiments on the different times rendered by different stimuli were corroborated by Wilhelm Heinrich von Wittich, *Archives de Pflüger* 2 (1869), 329 and "Durée de la persistance des sensations de tact dans les différentes régions du corps," *Travaux du laboratoire de M. Marey* 4 (1880).

finding the speed of nervous transmission with reaction time experiments. "In redoing more carefully the experiments" he "became assured that they did not have a sufficient precision."[43] He urged readers to look at "the extreme inequalities of personal reactions by studying the trace reproduced in Marey's book," referring to graph 69, a controversial graph representing the time taken by an observer to react to a stimulus.[44]

This trace was the result of experiments that Richet "did with M. Bloch" in the late 1870s.[45] It proved to him that "unfortunately the psychical acts involved in responding to a stimulus are too complex for the precise determination of the speed of nerve transmission."[46] The collaboration also convinced him that "a result could not be obtained with those methods, since one knows how averaging when the minimum values differ so much from the maximum is detestable in physiology."[47] He admitted results "differ greatly according to the experimenter."[48]

Marey's attempted rapprochement between Bloch and Richet backfired: now serious doubt hung over his previous claims.[49] The best measurements of reaction time in humans came only from the contentious results of Bloch and Richet and from their controversial graph.[50] Marey followed with an apostasy, acknowledging the imprecision and inconstancy of reaction time experiments for measuring the speed of sensory transmission. He criticized the method of differential nerve lengths, which he had once defended.[51]

In the second edition of his *Elements of Physiological Psychology,* Wundt cited Bloch's work as proof that the speed of sensory transmission could not be found through reaction time experiments because of great variations due to exercise, fatigue, and the place of stimulation.[52] More importantly,

43. Charles Richet, *Physiologie des muscles et des nerfs: Leçons professées à la Faculté de médecine en 1881* (Paris: Germer Baillière et cie, 1882), 583.

44. Charles Richet, "De la durée des actes psychiques élémentaires," *Revue philosophique de la France et de l'étranger, dirigée par Th. Ribot* 6 (1878): 396.

45. It was described in ibid.

46. Richet, *Physiologie des muscles et des nerfs: Leçons professées à la Faculté de médecine en 1881*, 582.

47. Ibid., 583.

48. Ibid., 581.

49. While Richet became an active member at the Institut Marey, Bloch did not. See Procès verbaux des réunions de commissions à l'Institut Marey à partir de 1902 et après la mort de Marey, jusqu'à 1959, Collège de France, Box CXII.

50. Richet, *Physiologie des muscles et des nerfs: Leçons professées à la Faculté de médecine en 1881*, 584 n. 1. When reminiscing about his experiments with Marey, Richet remembered how his graphs were never up to Marey's standards. Charles Richet, "Le professeur Charles Richet, autobiographie recueillie par le Dr. Pierre Maurel," in *Les Biographies médicales, notes pour servir à l'histoire de la médecine et des grands médecins* (Paris: J.-B. Baillière et fils, 1932), 146.

51. When it came to Hirsch's and Donders's work he remained unrepentant. Without acknowledging any controversy, he stood solidly behind their work.

52. Wilhelm Wundt, *Éléments de psychologie physiologique*, trans. Élie Rouvier, 2 vols. (Paris: Germer Baillière et cie, 1886), 2:254, 57.

Wilhelm Wundt, known for instituting reaction time experiments in his Leipzig laboratory, answered the question, "Are the individual differences which still persist . . . related to the temperament or to other particularities inherent to observers?" by saying that it "has not been answered by anyone," not even from the "examination of subjects affected by nervous or mental diseases."[53]

The crisis afflicted the graphic method in general, culminating in a contentious crusade against it in the Académie de médecine. One researcher fought against the "pretension that the graphic method should be the starting point of all exact and certain science."[54] This opponent of Marey argued that physiological phenomena, including the horse's gallop, could be grasped in a better way without complicated "self-registering" apparatuses. Traces, he argued, simply lacked "imitative harmony."[55] Furthermore, interpreting the graphs was as contentious as observing a galloping horse: "In examining with attention the traces, one confronts to similar degree the same difficulties that when in the presence of an animal which trots or gallops. . . . These lines tell you nothing."[56] Marey was accused of ignoring the fact that graphic traces themselves had to be "read," and that this introduced some of the same challenges of direct observations: "The registering apparatus does nothing but to inscribe undulating lines that fall on our senses; but once it comes to interpreting the traces, the graphic method has no more certitude than direct observation."[57] His contradictor contested the supposed universality of their language: "If everyone can produce these traces, not everyone is capable of interpreting them."[58] After listening to these accusations, even some of the graphic method's most hardened supporters came to doubt the claim that they constituted a universal, easily interpretable language. An observer concluded: "In the sciences of observation, all instruments, no

53. Ibid., 257.

54. "Académie de médecine. Séance du 9 juillet 1878.—Présidence de M. Richet," *Gazette médicale de Paris* 7 (1878): 840.

55. For the criticisms of Gabriel Colin against Marey see Gabriel Colin, *Traité de physiologie comparée des animaux considérée dans les rapports avec les sciences naturelles, la médecine, la zootechnie et l'économie rurale*, 3rd ed., 2 vols. (Paris: J.-B. Baillière, 1886), 1:40. Although in the third edition he included Marey's work, he still kept the illustrations of his first edition, even of a horse in the controversial *ventre à terre* position. Compare Gabriel Colin, *Traité de physiologie comparée des animaux domestiques*, 2 vols. (Paris: J.-B. Baillière, 1854), 1:332, fig. 35, and *Traité de physiologie comparée des animaux considérée dans les rapports avec les sciences naturelles, la médecine, la zootechnie et l'économie rurale*, 1:483, fig. 78.

56. "Observations à l'occasion du procès-verbal. II. Méthode graphique," *Bulletin de l'Académie de médecine* 7 (1878): 724.

57. "Communications 1. Sur l'importance au point de vue médical des signes extérieurs des fonctions de la vie," *Bulletin de l'Académie de médecine* 7 (1878): 622.

58. Ibid.

mater how simple or complicated, are aids . . . that speak a special language. Before using them, one must strive to learn their language."[59]

Errancy Effects

A few years after Richet and Bloch's investigations, reaction time experiments were once again placed on trial by Albert René, an assistant working for Henri Beaunis (who would later head the first laboratory of experimental psychology in France).[60] René tried to improve on previous reaction time research by abandoning the "the method of differential nerve lengths." Acknowledging that nerves from different parts of the body could not be compared against each other, he left this method aside and measured the speed of thought by comparing reflex reactions to voluntary ones.[61] René gave new numbers for cerebral acts, but cautioned: "there is no absolute and constant value for the speed of nerve transmission." Furthermore, these numbers varied greatly according to the intensity of the stimulus, the type of stimulus, normal and pathological subjects, and whether one was experimenting on students of medicine and doctors or on adolescents. Setting the tone for a century of research on reaction time, he concluded that the only place where these types of investigations "could be useful" was in their applications to pathology. René continued his investigations there.[62]

Beaunis took on a challenge greater than that of René. He experimented on a particularly recalcitrant sense, the reaction time of smell.[63] His summer

---

59. Gavarret, "Observations à l'occasion du procès-verbal. III. Méthode graphique," *Bulletin de l'Académie de médecine* 7 (1878): 760–61.

60. Albert René, "Étude expérimentale sur la vitesse de transmission nerveuse chez l'homme: Durée d'un acte cérébral et d'un acte réflexe, vitesse sensitive, vitesse motrice," *Gazette des hôpitaux* 55 (1882).

61. For a "medium" intensity "the time necessary for sensory perception plus the time necessary to will, transmit and effect a movement" was .212 seconds. He found the "approximate" time of "elementary" cerebral acts by subtracting the time of reflex action from the total time of reaction, furnishing .035 to .032 seconds. To find the time of sensory transmission René subtracted the time of reaction to an auditory stimulus (.179) from the time of reaction for a touch stimulus (.212). To obtain the speed of sensory transmission he divided this number over a nerve length of 95 centimeters, rendering a speed of 28 meters per second. Richet later used this same process in his experiment with "normal" frogs.

62. René, "Étude expérimentale sur la vitesse de transmission nerveuse chez l'homme: Durée d'un acte cérébral et d'un acte réflexe, vitesse sensitive, vitesse motrice." René found a delay in the reaction time of persons suffering from motor ataxia, while in persons suffering from general paralysis the delay was absent and the reaction at times accelerated.

63. His investigations appeared in the *Revue médicale de l'est* (February and March 1883) and in the *Gazette médicale de Paris* (10 February 1883). An extract was published in Henri Beaunis, "Sur le temps de réaction des sensations olfactives," *Comptes rendus des séances de l'Académie des sciences* 96 (1883). They were preceded by the work of G. Buccola, "Sulla durata delle percezioni olfattive," *Archivo Italiano per le malattie nervose* 6 (1882). On the reaction time of smell, also see W. Moldenhauer, *Philosophische Studien*, ed. Wilhelm Wundt, no. 4 (1883) and a review of it in "Revue des périodiques étrangers: Philosophische Studien, Hgg. von W. Wundt. 1883. 4e Heft.," *Revue philosophique de la France et de l'étranger, dirigée par Th. Ribot* 15 (1883). For taste, see Wundt, *Éléments de psychologie physiologique*, 256. Taste and smell had not been investigated by Bloch not only because they were "too complicated," but because they required

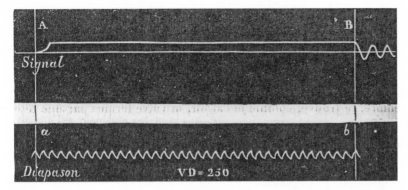

Figure 3.2. Graph of the time taken for a person to react to a stimulus. Rhetorically, the graphics published by René were different from Bloch's and Richet's. By showing only a single reaction at a time, they minimized the problem of variations among and within individuals. From Albert René, "Étude expérimentale sur la vitesse de transmission nerveuse chez l'homme: Durée d'un acte cérébral et d'un acte réflexe, vitesse sensitive, vitesse motrice," *Gazette des hôpitaux* 55 (1882), p. 276.

of 1883 was filled with experiments on himself and on a number of students. He acknowledged that times of reaction varied according to the intensity of the stimulus.[64] This admission was accompanied by an acknowledgment of another complicated question, that of comparing different senses and stimuli. What could physiologists and astronomers possibly mean when they claimed the ear was faster than the eye? These comparisons depended on different types of stimuli (drum beats versus light impulses, for example). How could these be compared? To complicate things even more, one day, when he was feeling unwell, he noticed his reaction time increase dramatically.[65] Eugène Gley, who had studied with Beaunis at Nancy and who worked in Marey's laboratory, reviewed Beaunis's work. After offering courteous remarks, he concluded: "However, the results of the experiments done on the duration of visual sensation and of the sense of taste can still not be considered as definitive."[66]

---

voluntary action, such as breathing or swallowing. All voluntary actions, according to Bloch, were out of bounds for physiological investigations. Binet and Jacques Passy continued to investigate the reaction time of smell and its smallest noticeable differences. Passy determined the perceptible minimum of vanilla to be 0.000005 grams for a liter of air. Bloch, "Psychologie: La Vitesse comparative des sensations," 586. Binet cited Passy's work in Alfred Binet, "Travaux du laboratoire de psychologie physiologique (Hautes Études): La Perception de la durée dans les réactions simples," *Revue philosophique de la France et de l'étranger, dirigée par Th. Ribot* 33 (1892): 650.

64. Henri Beaunis, "Sur la comparaison du temps de réaction des différentes sensations," *Revue philosophique de la France et de l'étranger, dirigée par Th. Ribot* 15 (1883).

65. Henri Beaunis, "Société de psychologie physiologique: I.—Influence de la durée de l'expectation sur le temps de réaction des sensations visuelles," *Revue philosophique de la France et de l'étranger, dirigée par Th. Ribot* 20 (1885): 332.

66. Eugène Gley, "Notices bibliographiques," *Revue philosophique de la France et de l'étranger, dirigée par Th. Ribot* 15 (1883).

Interrogating the Subject

Reaction time occupied other investigators at Nancy. Augustin Charpentier, professor of physics at the University of Nancy, joined forces with Hippolyte Bernheim, an important influence on Freud who started questioning the methods of Jean-Martin Charcot. Admitting that "at first glance . . . no serious conclusion could be obtained from such divergent results," he chose a strategy that was altogether different from that of his predecessors.[67]

Charpentier and Bernheim, in collaboration, noticed how their times of reaction were affected by attention, exercise, and fatigue; by the retinal point that received the stimulus; and by a "general state of nervousness." Charpentier blamed divergent numbers on a "general disposition of the brain," and, although he cautioned that this state was "difficult to analyze," he continued to experiment. On one occasion Bernheim was conversing with Charpentier during the experiment about a certain "topic of pathological anatomy," when his reaction time increased significantly. Charpentier decided to run further tests on this matter, asking Bernheim to recite "lessons on pneumonia lesions." Reaction times continued to increase. Attention, Charpentier learned, "is not always voluntary" and "despite the best efforts of the will, it is dominated in part by ideas or other preoccupations."[68]

While the dependency of reaction time on these factors was frequently noted, Charpentier offered a novel solution. The disorder of the results could be diminished if the subject under experimentation described his state of mind. The subject's own perception of his "general disposition," Charpentier ventured, might shed light on otherwise confusing and sundry numbers. According to the Nancy doctor, the "subject had a clear conscience of [this state]: it is, he says, well or badly disposed."[69] His insight arose from listening to his talkative experimental subject, the controversial psychologist Bernheim.

During these experiments Charpentier tried to overcome some objections raised against reaction time experiments. What differentiated his experiments from previous ones was that the experience of the observer, instead of being discounted, validated the experiment. By paying attention to the confessions of the subject, some order could be imposed on the otherwise erratic results. Charpentier explained: "There are days when we say before beginning the experiment, I feel less disposed to react, and, in fact, the reactions were slower. The general state of the nervous system is thus

---

67. Augustin Charpentier, "Recherches sur la vitesse des réactions d'origine rétinienne," *Archives de physiologie normale et pathologique, publiées par MM. Brown-Séquard, Charcot, Vulpian* 15 (1883): 619.
68. Ibid., 622–23.
69. Ibid., 620.

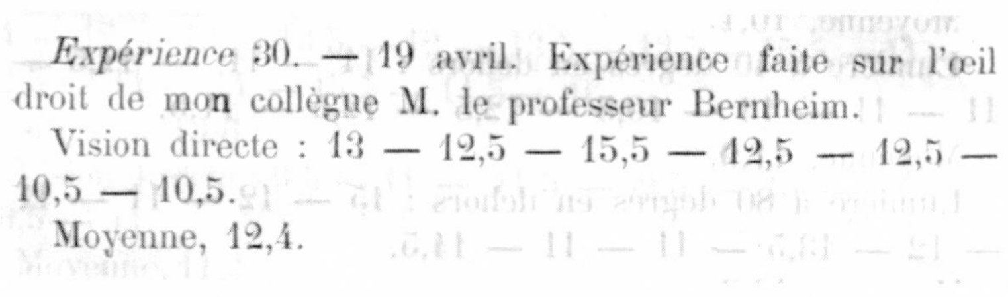

Expérience 30. — 19 avril. Expérience faite sur l'œil
droit de mon collègue M. le professeur Bernheim.
    Vision directe : 13 — 12,5 — 15,5 — 12,5 — 12,5 —
10,5 — 10,5.
    Moyenne, 12,4.

Figure 3.3. Reaction time experiments performed on Bernheim by Charpentier. The results are
in hundredths of a second. From Augustin Charpentier, "Recherches sur la vitesse des réactions
d'origine rétinienne," *Archives de physiologie normale et pathologique, publiées par MM. Brown-
Séquard, Charcot, Vulpian* 15 (1883): 599–635, on p. 618.

an important factor that, even if it escapes physiological analysis, it is perfectly appreciated by our conscience."[70] According to him, the information
obtained from the "subject's conscience" was relevant. By letting the subject
speak, Charpentier obtained new information, but cautiously remarked:
"Voilà, this is all I can precisely say on this topic, and it is not much."

    Charpentier valued the subject's own description of his internal states,
clearly borrowing from a radically different experimental tradition based
on introspection and interrogation. But these techniques were not the solution after which most experimenters searched.[71] Because they required an
additional confession from the experimental subject, these graphic experiments could not fulfill Marey's goal of providing a representation that was
"superior to all other forms of expression."[72]

Discredit

By the last decades of the century, disagreement on the interpretation of
reaction time experiments was rampant. Ribot lamented how research was

70. Ibid., 622.
71. Both Delboeuf and Wundt found interrogation methods to be scientifically suspect. They discarded
them as inexact and subject to errors. Delboeuf insisted on the "incommunicability of internal phenomena": "Between the experimenter and the subject on whom he would want to experiment there exists a . . .
considerable barrier from a quantitative perspective. In effect, internal phenomena, sensation, pain, fatigue, well-being, can have many varied intensities, but the intimate sense can never succeed at comparing these intensities other than in a coarse manner, and their appreciation does not go beyond superiority
or inferiority.—My pain is strong, very strong, minimal, large, very large as large as last night's,—voilà,
how all patients express themselves; they cannot use any other form of measurement." Joseph Delboeuf,
"La Mesure des sensations: Réponses à propos du logarithme des sensations," *Revue scientifique* 8 (1875):
1016. Similar comments appear in Joseph Delboeuf, "Étude psychophysique: Recherches théoriques et expérimentales," *Mémoires couronnés et autres mémoires publiés par l'Académie royale des sciences, des lettres
et des beaux-arts de Belgique* 23 (1873): 4. For Wundt's critique of interrogation, see Wundt, "Über Aus-
frageexperimente und über die Methoden zur Psychologie des Denkens," *Psychologische Studien* (1907)
3:301–60, on pp. 329–339, and his comment on suggestion on pp. 339–40.
72. Marey, *La Méthode graphique dans les sciences expérimentales et principalement en physiologie et en
médecine,* iii.

"dispersed in a great number of books, memoirs and *recueils* published in different languages." Even worse, "the same experiments have been undertaken by many scientists who are frequently in disaccord with each other, either with respect to the results produced, or their interpretation."[73] In order to minimize the confusion, he tried to make experimental languages comparable. But Ribot acknowledged that some of the problems facing experimental psychology were not only due to translation difficulties. He admitted, for example, that reaction time "does not lend itself" to investigations on the duration of mental acts.[74]

From the late 1870s to the late 1880s, the experimental system consisting of stimulus, subject, and inscription device risked falling into complete discredit in physiology. In 1884 Bloch repeated his 1875 experiments using a new method, so as to corroborate his previous results. He had "at length and explicitly demonstrated that those experiments [involving voluntary responses] are fundamentally defective."[75] He could not have been clearer: "Voluntary responses contain a psychic element whose duration, and manner of being, is unknown: the transformation of perception into volition. This element varies from one sensation to the next, without us knowing why. Habit modifies it, exercise accelerates it; it does not have the stability which is indispensable for numerical investigation, and once it emerges, the physiological experiment no longer has any value."[76] Bloch traced the cause of the problem to the astronomer Charles Wolf's classic 1866 work on the personal equation. Astronomers, he argued, had introduced a "detrimental confusion" into personal equation research. The confusion resided in that the term "personal equation" encompassed two distinct phenomena. One was "purely sensorial" (determined by the method of tactile persistence), and the other one was "psychic" involving volition. The reaction of a living subject introduced an indeterministic "psychical" aspect. Only by creating an experiment where "the element of volition is never in play" and "things are purely sensorial" could one study physiological man. By not distinguishing sensation from reaction, astronomers had "introduced a damaging error into science."[77]

73. Théodule Ribot, "Analyses: G. Buccola. La legge del tempo nei fenomeni del pensiero: saggio di psicologia sperimentale," *Revue philosophique de la France et de l'étranger, dirigée par Th. Ribot* 16 (1883): 421.
74. Ibid., 422.
75. A.-M. Bloch, "Expériences nouvelles sur la vitesse du courant nerveux sensitif chez l'homme," *Journal de l'anatomie et de la physiologie normales et pathologiques de l'homme et des animaux* 20 (1884): 289. The results of these investigations also appeared in A.-M. Bloch, "Expériences nouvelles sur la vitesse du courant nerveux sensitif chez l'homme," *Comptes rendus des séances et mémoires de la Société de biologie* 1 (1884).
76. Bloch, "Expériences sur la vitesse relative des transmissions visuelles, auditives, et tactiles," 26.
77. Ibid.

In the third edition of his popular physiology textbook (1888) Beaunis recounted how reaction time was influenced by attention, exercise, the intensity of sensation, fatigue, certain substances, and even age, sex, race, and state of health.[78] Astronomers, psychologists, and physiologists alike accepted the dependency of the personal equation on a myriad of factors.

Psychological Reactions

Richet and Bloch's collaboration convinced scientists of the difficulties in finding the speed of nervous transmission and of the instability of reaction time. The agreement of the scientific community, however, ended there. Scientists still disagreed on a matter of more profound consequences: the speed of thought. While Bloch interpreted the variations in reaction time measurements (as shown in graph 69) as indicative of a faulty experimental method, Richet disagreed. For him, the graph showed the possibility of measuring the speed of thought in an exact and quantitative manner. Given that the speed of sensitive nervous transmission was nearly impossible to find with reaction time experiments, the more interesting "mental" phenomena still remained to be studied. Increasingly after his collaborative work with Bloch, Richet claimed that reaction time experiments measured psychic actions. These were experiments by which "one measured the duration of psychic acts, that is to say cerebral work, that which astronomers have for a long time referred to as the personal equation."[79] Although he had no choice but to admit, with Bloch, the unexplainable anomaly that a subject reacted more quickly to a stimulus on the hand than on the face, he ignored all these contradictions. He continued to experiment in Marey's laboratory finding results roughly agreeing with those of the so-called German school associated with Wundt.

Ribot contested Bloch's division of the human subject into two components: one passive, physiological, and measurable; and the other psychological, active, and unquantifiable. Bloch's experiments on tactile sensory persistence, he insisted, were not any more trustworthy than reaction time experiments. They did not fulfill Bloch's own goal of purifying sensation by separating it from mental processes. Sensory persistence, he argued, also had a mental component. This was the time for "apperception," that is, the time taken by the brain to recognize different impressions. Through

---

78. The discrepancies were considerable, ranging from 30 meters per second (Schelske) to 132 meters per second (Bloch). Henri Beaunis, *Nouveaux éléments de physiologie humaine comprenant les principes de la physiologie comparée et de la physiologie générale*, 2 vols. (Paris: J.-B. Baillière et fils, 1888), 1:648–50.

79. Richet, *Physiologie des muscles et des nerfs: Leçons professées à la Faculté de médecine en 1881*, 865.

the concept of apperception, Ribot attacked Bloch's claim that his method bypassed the brain and studied "pure sensibility." He thus attributed a mental, psychological quality even to the passive subject of Bloch's experiments.[80] Psychological, intellectual, and individual factors surfaced in both active and passive observations. Psychology was permanently lodged in the body—regardless of whether scientists dealt with sensory or motor responses, which in any case he considered to be inextricably linked.

Richet used this same argument against Bloch. Bloch's method of testing the feeling of simultaneity in passive subjects depended "on a psychic element whose reality we ignore completely and which is difficult to appreciate"—namely, the perception of stimuli applied at different times and on different body parts.[81] He discredited Bloch's number for the speed of sensory transmission and continued to experiment with reaction time.[82] Wundt followed this same line of reasoning, claiming that in Bloch's experiments "the psychological influences were not completely eliminated, as the author believed."[83]

Instead of ending controversy, Richet and Bloch's work in Marey's laboratory brought personal equation experiments to a new level of scrutiny. Even though they failed to forge consensus, experimentation on reaction times increased. These experiments continued to point toward new possibilities, in particular, to the possibility of treating individual intellectual factors in exact, numerical terms.

From Visual to Direct Reading Instruments

When new instantaneous photographs showed horses, humans, and even birds in unrecognizable postures, many scientists believed that these photographs might overcome the limitations of preexisting scientific imaging techniques.[84] When Bloch first saw them, they taught him how the visual

---

80. Théodule Ribot, *La Psychologie allemande contemporaine (école expérimentale)* (Paris: Libraire Germer Baillière et Cie, 1879), 321–22.

81. Richet, *Recherches expérimentales et cliniques sur la sensibilité*, 52. In his later work he repeated a criticism already present in his thesis of 1877: "In effect, even [in Bloch's experiments] a psychical element persists: the perception of the simultaneity of two stimuli." Richet, *Physiologie des muscles et des nerfs: Leçons professées à la Faculté de médecine en 1881*, 585.

82. "As a summary, for the speed of sensitive nerves, despite the incertitude of our present notions, it seems that the value with the most verisimilitude is 50 to 60 meters per second." Richet, *Physiologie des muscles et des nerfs: Leçons professées à la Faculté de médecine en 1881*, 583.

83. Wundt, *Éléments de psychologie physiologique*, 250 n. 1.

84. For a history of "direct reading" instruments, see Graeme J. N. Gooday, *The Morals of Measurement: Accuracy, Irony and Trust in Late Victorian Electrical Practice* (Cambridge: Cambridge University Press, 2004).

sense functioned as a magic lantern that constantly fused discontinuous reality and provided a sense of movement:

> It is the instinctive education furnished by visual persistence that makes the postures of passersby in instant photographs seem so bizarre. In our sensorium, every move-ment is tied to that which precedes it and that which will follow. We do not see the galloping horse, as it is, but as it was and as it will be in a fraction of a second. So it is this continuous, unconscious synthesis that takes away the possibility of accepting as real the truth itself of things, as when it is shown to us by a photograph.[85]

As he had once done with the graphic method, Marey now started preach-ing about how *photographs* could be used to reveal a "natural expression which is not that of ordinary language: the latter is too slow and to impre-cise to describe clearly, with their complexity and variability, the different acts of life."[86] Yet even with photographs, scientists continued to disagree about the nature of the delay they revealed. How were visual lag times related to reaction time delays? How could these lag times be disentangled from those noticed in reactions? Which part of reaction lag times were sen-sorial, which were due to nerve transmissions, which to muscular inertias, and which were mental?

In light of these debates, reaction time experiments moved beyond labo-ratories with the creation of an new chronoscope built by Arsène d'Arsonval, a student of Brown-Séquard and assistant to Claude Bernard (from 1874 to 1878) who would become the founder of biophysics in France.[87] In 1885 D'Arsonval introduced a distinctive chronometer, "destined" to measure the speed of sensory transmissions. "Well-known" soon after its invention, it would be foundational for French experimental psychology.[88]

The instrument's novel design helped free reaction time experiments from their exclusive moorings in the physiology laboratory and adapt them

85. Bloch, "Psychologie: La Vitesse comparative des sensations," 588.
86. Étienne-Jules Marey, "La Photochronographie et ses applications à l'analyse des phénomènes physi-ologiques," *Archives de physiologie normale et pathologique, publiées par MM. Brown-Séquard, Charcot, Vulpian* 1 (1889): 516.
87. D'Arsonval considered his research a continuation of Brown-Séquard's early work. In the presentation of his famous chronometer he explained how it "was especially created for the physiological and patholog-ical research on the speed of nervous transmission, research undertaken in the past by Brown-Séquard, and continued with my collaboration." Arsène d'Arsonval, "Chronomètre à embrayage magnétique pour la mesure directe des phénomènes de courte durée (de une seconde a 1/500e de seconde)," *Comptes rendus des séances et mémoires de la Société de biologie* 3 (1886): 236.
88. Brown-Séquard, "Bibliographie: Recherches expérimentales sur la durée des processus psychiques les plus simples et sur la vitesse des courants nerveux à l'état normal et à l'état pathologique, par le Dr. A. Rémond (de Metz)," *Archives de physiologie normale et pathologique, publiées par MM. Brown-Séquard, Charcot, Vulpian* 1 (1889).

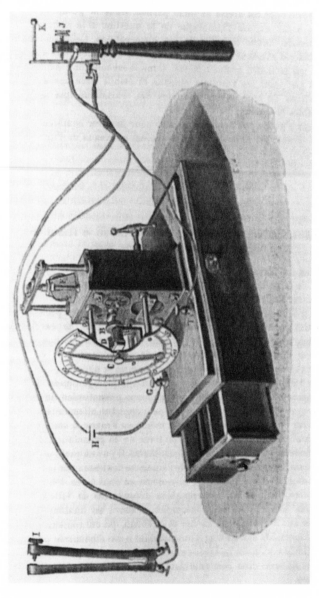

Figure 3.4. The D'Arsonval chronometer. From Brown-Séquard, "Bibliographie: Recherches expérimentales sur la durée des processus psychiques les plus simples et sur la vitesse des courants nerveux à l'état normal et à l'état pathologique, par le Dr. A. Rémond (de Metz)," *Archives de physiologie normale et pathologique, publiées par MM. Brown-Séquard, Charcot, Vulpian* 1 (1889): 612–14. on p. 613.

to the clinic and points beyond. D'Arsonval was intimately familiar with Marey's graphic method. Although he considered it "very precise," it still "required tools and multiple manipulations which often make its use difficult outside of the laboratory, especially in the clinic."[89] For this reason, he decided to transform graphic techniques into direct reading machines "particularly created for measuring visual, auditory, and tactile psychomotor reaction times."[90] His chronometer was "of very small dimensions" and "could easily be carried in a pocket for clinical experiences."[91] It made reaction time technology nearly wearable. By taking mental chronometry "outside of the laboratory" he hoped to "help clinical nervous diseases." The instrument, D'Arsonval rightly surmised, might stimulate research. With it, instead of having to interpret graphic traces, scientists could simply read a number from a numerical scale.

The D'Arsonval chronometer was extraordinarily successful. Its use in clinical contexts decidedly changed the focus of reaction time research. Inspired by "the apparition of M. d'Arsonval's instrument," Antoine Rémond, a student in Nancy, used it to measure the reaction time of "hysterical" subjects and compare them to healthy ones.[92] Brown-Séquard wrote a glowing review of Rémond's thesis, arguing that he had found "approximate values" for the speed of nerve transmission. Furthermore, Rémond had succeeded in measuring "the most simple" mental operations.[93]

D'Arsonval dedicated his first scientific paper to the problem of the personal equation. For more than half a decade thereafter, he considered the elucidation and elimination of the personal equation as the central motivation behind his work.[94] This goal represented the ultimate achievement of

89. D'Arsonval, "Chronomètre à embrayage magnétique pour la mesure directe des phénomènes de courte durée (de une seconde à 1/500e de seconde)," 235. D'Arsonval worked with Marey after having a personal fight with his mentor Brown-Séquard.

90. Léon Binet, "A. d'Arsonval, physiologiste et médecin," in Cérémonie commémorative en l'honneur du centenaire de la naissance du professeur Arsène d'Arsonval 1851–1940, ed. L'Union des associations scientifiques et industrielles françaises (Paris: Draeger, 1952). The times in normal man were found to be .19 seconds for visual and .14 seconds for auditory and tactile reactions.

91. D'Arsonval, "Chronomètre à embrayage magnétique pour la mesure directe des phénomènes de courte durée (de une seconde à 1/500e de seconde)," 236. The instrument was built by Verdin.

92. A. Rémond, Recherches expérimentales sur la durée des actes psychiques les plus simples et sur la vitesse des courants nerveux à l'état normal et à l'état pathologique (Paris: Octave Doin, 1888).

93. Brown-Séquard, "Bibliographie: Recherches expérimentales sur la durée des processus psychiques les plus simples et sur la vitesse des courants nerveux à l'état normal et à l'état pathologique, par le Dr. A. Rémond (de Metz)." Another review appeared in "Psychological Literature: Recherches expérimentales sur la durée des actes psychiques les plus simples et sur la vitesse des courants nerveux à l'état pathologique. A. Rémond. Paris: Octave Doin. 1888. pp. 135," American Journal of Psychology 2 (1888).

94. D'Arsonval's first scientific paper, "The Personal Equation of the Astronomers," was partially republished in Léon Delhoume, De Claude Bernard à D'Arsonval (Paris: J.-B. Baillière & fils, 1939), 220–26. It first appeared in the Revue médicale de Limoges 4, no. 4 (April 1873). While this work originally comprised two parts, only the first one was published.

scientific progress. Instead of framing the medical revolution in terms of the monumental germ theory of disease advanced by Louis Pasteur or the antiseptic revolution associated with Joseph Lister, D'Arsonval charted the progress of science by the gradual mastery over the personal equation. In a speech given to the Collège de France commemorating the hundred-year anniversary of the death of the famous physician René Laënnec, he framed the entire modern revolution in medicine in those terms: "The purely sensorial information [collected during diagnosis] was affected by that which the *astronomers call the personal equation* of the observer, which varied according to different sensibilities and momentary judgments."[95] The first step in solving this problem of variability, according to D'Arsonval, was the revolutionary replacement of direct auscultation with the use of the stethoscope. Later instrumental changes were to reduce even further the personal equation and fuel the progress of modern medicine.

Stick Figures

But by any count, the lack of agreement surrounding reaction time experiments promoted rather than curtailed research. For years scientists had tried to eliminate individual differences in reaction time experiments. But by the 1890s they experimented with them in order to intensify these differences. The psychologist Charles Féré gladly abandoned all hope of finding an exact number for the speed of sensory transmission since "reaction times are subject to considerable variations connected to modifications in circulation and nutrition, which lead us far from the precise formulas of German psychometry."[96] Admitting that reaction time "varies according to the subject," he now added something else: "these personal variations are extraor-

95. This progress occurred when "tactile sensations to detect fever" were replaced by the "constant precision of a thermometer; the feeling of the pulse with the impersonal trace of the sphygmograph and of the electro-cardiograph; the organoleptic character of moods with physico-chemical analysis; powerless vision by x rays; the mystery of the nervous system with electro-diagnosis and chronaxia; the subjective with the objective." Cited in Louis Chauvois, *D'Arsonval: Soixante-cinq ans à travers la science* (Paris: J. Oliven, 1937), 425–26. Italics mine.

96. Charles Féré, "Note sur le temps d'association, sur les conditions qui le font varier et sur quelques conséquences de ses variations," *Comptes rendus des séances et mémoires de la Société de biologie* 2 (1890). For these investigations Féré used the D'Arsonval chronometer, but at other times he used Marey's graphic method. For the relation between nutrition, attention, work, and reaction times, see "L'Énergie et la vitesse des mouvements volontaires" *Revue philosophique* (July 1889). A review of this work appeared in "Psychological Literature: L'Énergie et la vitesse des mouvements volontaires. Ch. Féré. Revue philosophique. Juillet, 1889," *American Journal of Psychology* 2 (1889). Also on reaction time see Charles Féré, "Note sur la physiologie de l'attention," *Comptes rendus des séances et mémoires de la Société de biologie* 2 (1890), and "Le Travail et le temps de réaction," *Comptes rendus des séances et mémoires de la Société de biologie* 4 (1892).

dinarily interesting."[97] In his *Pathology of Emotions* (1892), he found that the average reaction time in formulating word associations was .7 seconds in men and .83 seconds in women. Variations within individuals, among individuals, and among groups were no longer errors, but valuable results. The names of subjects under investigation frequently figured prominently in the results.

For years, Bloch held his ground, continuing to argue against reaction time experiments. His patience in the face of their increasing popularity was tried, once again, in 1891, when he criticized these experiments in a debate with a doctor from the physiological laboratory of the Bucharest medical faculty.[98] The doctor's results, appearing in the prestigious *Comptes rendus* of the Société de biologie, prompted a vehement response from Bloch. He compared the reaction time of persons suffering from sensory and motor derangements due to an inflammation of the spinal cord, to the numbers found by Marey and René in normal individuals and found that they were significantly lower. The doctor from Bucharest vigorously campaigned for the use of reaction time measurements in diagnosis. Bloch's response invoked the criticisms he had published more that a decade ago. Studying the speed of sensory transmission through reaction time experiments, he stressed, was "totally opposed to scientific truth." The problem, he added, was not technical. Perfecting chronographic instruments (as d'Arsonval did) was irrelevant "because the foundations of the method themselves are wrong."[99]

The doctor from Bucharest answered Bloch's attacks by revising his earlier claims. He changed his views with respect to the speed of sensory transmission and agreed with Bloch that physiologists were unsure about how the results of reaction time experiments fit with his findings. He, however, continued to claim that he had found significant differences in the time of reaction of normal and pathological subjects. One thing was clear: reaction time measured something. For the doctor, this was more than enough. Even if scientists still did not know what reaction times measured, they could still be used. While Bloch continued to insist that the speed of sensory transmission could not be found through reaction experiments, after this debate he

97. Féré, "Note sur le temps d'association, sur les conditions qui le font varier et sur quelques conséquences de ses variations," 174.

98. Grigorescu used a D'Arsonval chronometer for his experiments. G. Grigorescu, "Application du chronomètre électrique de M. d'Arsonval au diagnostic des myélites," *Comptes rendus des séances et mémoires de la Société de biologie* 3 (1891): 364.

99. A.-M. Bloch, "Note à propos de la communication faite par M. G. Gregorescu [sic], le 16 mai," *Comptes rendus des séances et mémoires de la Société de biologie* 3 (1891): 379, and "Deuxième note relative aux expériences de M. Grigorescu," *Comptes rendus des séances et mémoires de la Société de biologie* 3 (1891).

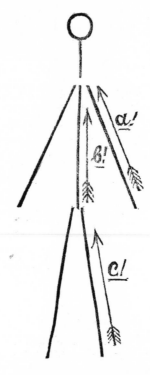

Schéma de la conduction
sensitive.

Figure 3.5. Diagram showing the different paths of nerve conduction. From G. Grigorescu, "Application du chronomètre électrique de M. d'Arsonval au diagnostic des myélites," *Comptes rendus des séances et mémoires de la Société de biologie* 3 (1891): 364–65, on p. 364. Courtesy of the Ernst Mayr Library, Museum of Comparative Zoology, Harvard University.

accepted that they could be "useful" for studying the diseases of the nervous system. The Romanian doctor compared his work to scientists who worked to establish longitude and length standards and who frequently debated about the status of their standards.

Even if these scientists did not know what a kilometer meant in absolute terms, sturdy networks of metrology increasingly permitted them to measure in kilometers. Similarly, even if physiologists did not know exactly what reaction times measured, they could still be used to determine "distances." Reaction time could function like the meter zealously guarded in the vaults of the Conservatoire des arts et métiers. It could be used as a standard, comparable to length standards: "Let me summarize this little discussion by an example. I have measured in kilometers the distance from Paris to Bucharest, and from Paris to Berlin, and, without figuring out a theory

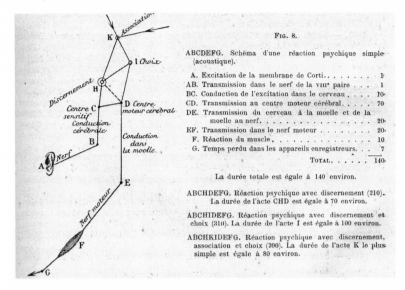

Figure 3.6. Schematic drawing of the stimulus-reaction process. From Charles Richet, "Cerveau," in *Dictionnaire de physiologie*, edited by Charles Richet, 1–36 (Paris: Alcan, 1898), p. 33. Courtesy of the Ernst Mayr Library, Museum of Comparative Zoology, Harvard University.

of the kilometer on which everyone could agree, I have found a remarkable difference between these two distances. And so, this difference is real, even if it is only in kilometers."[100] The argument hinged on the claim that just as a real, measurable difference existed between Paris and Bucharest, an analogous difference existed between fast and slow reactions. Over the next decades, the participants of the debate succeeded in defining differences in reaction time as sort of physiological "distance" and comparing it to other measurement categories, such as age, race, nationality, educational level, and state of health.

Reaction Networks

In 1898 Richet summarized the preexisting scholarship on reaction time, arguing for the central importance of the tenth of a second. He offered a diagram explaining the relation of reaction time to the speed of thought and published it in his famous *Dictionnaire de physiologie,* a ten-volume (1895–1928) classic reference text. The diagram detailed the various steps and times needed to react to auditory stimuli.

100. G. Grigorescu, "Réponse à la note de M. Bloch, relative à ma communication du 16 mai," *Comptes rendus des séances et mémoires de la Société de biologie* 3 (1891): 476–77.

Reaction time was comprised of two elements, one psychological and the other one physiological. Both added up "to the minimum of 100, that is to a *tenth of a second*." Richet noted that detractors to the interpretation of reaction time as comprised in part by the speed of thought were now few. But among them stood, most notably, Wundt: "It seems to me that this reasoning is fair and right, and I do not understand why Wundt (1886, 254) refuses to accept it."[101] Richet referred to the section in Wundt's book where he cited Bloch's research.

In one sense, the questions regarding reaction time experiments were part of a general discussion about what constituted a legitimate scientific experiment. Despite Richet's efforts, a single interpretation of reaction time experiments failed to coalesce. Scientists replaced the graphic method with direct reading instruments that could be used in different settings. They compared their numerical results to other standards, such as those of length. Their experiments were trailed by instrumental improvements, clinical applications, and promises of using them to enact great social revolutions, but not all the objections raised against them could be dispelled. One component of the scientific community opposed the view that an experiment that included a person's willful reaction to stimuli was valid. This group, however, was slowly left in the minority. Richet and others successfully changed the meaning of experimentation. In the process they legitimated experimental systems where the subject under experimentation was an accepted component within a system composed of keys, wires, and automatic inscription devices.

101. Richet, "Cerveau," 21.

CHAPTER 4

MOMENTS OF CONTACT

Almost constant and *VERY SIGNIFICANT* differences persist between different
observers, especially in estimating the time of . . . contacts.

EDMOND DUBOIS, science writer

In the 1860s the astronomer Charles Wolf built an artificial star machine that recreated, as closely as possible, the experience of observing transit stars. The machine used artificial stars that passed in front
of a telescope's sighting wires and automatically recorded the time of
their passage. When an observer saw the artificial star pass in front
of the wires and pressed a telegraphic key, the time as noted by the
observer was compared against the time automatically recorded by
the artificial star. In a publication that would become a "classic," Wolf
explained how when first using his instrument, he noticed that a difference of about three-tenths of a second existed between the real moment of the star's passage and his own appreciation of this moment,
and he decided to run further experiments. After using it systematically
for six months, from January to June 1864, he was able to reduce the
difference from three-tenths to one-tenth of a second, when his time

stabilized.[1] Once he had shortened his delay to a tenth of a second, he found that he could not reduce it any further: "the spirit, once it arrives at a certain limit, cannot go beyond it under any circumstances."[2] At this time scale, Wolf found an "indivisible moment of time."[3]

Wolf associated this value with the phenomenon of visual persistence that had fascinated scientists for centuries. Visual persistence effects had been known since antiquity, but the first attempts to measure the duration of persistence dated from the late eighteenth century. Yet even by the middle of the nineteenth century considerable debate surrounded measurements of its duration. Wolf undertook investigations on the persistence of visual impressions anew. He lamented how previous investigations (he cited those of Isaac Newton, Patrice D'Arcy, and Joseph Plateau) had all used different methods and arrived at different conclusions. In contrast, Wolf standardized these experiments by setting certain parameters, making sure, for example, that he tested the same retinal point and that he controlled the intensity of the light source. He concluded that, under certain fixed conditions, effects of fusion appeared when a light source undergoing "10 turns per second" illuminated the same retinal point.[4] Wolf's work immediately became a classic.[5] Astronomers all over the world frequently referred to it, contested it, and built more artificial transit machines to investigate the allegedly irreducible tenth-of-a-second delays.[6]

The Transit of Venus

A few years after his initial investigations, Wolf started the construction of a newer machine. While the previous instrument had been used to analyze observations that were performed daily at observatories to determine time and longitude, this new machine would be used in preparation for the "astronomical event of the century": the 1874 transit of Venus across the sun.

---

1. Wolf's artificial machine was used by P. Tacchini, "Sulla equazione personale," *Rivista sicula di scienze, letteratura ed arti* 2 (1869): 382. Cited by Gabriele Buccola, *La legge del tempo nei fenomeni del pensiero: Saggio di psicologia sperimentale,* Biblioteca Scientifica Internazionale (Milan: Fratelli Dumolard, 1883), 37:170.
2. Charles Wolf, "Recherches sur l'équation personnelle dans les observations de passages, sa détermination absolue, ses lois et son origine," *Annales de l'Observatoire impérial de Paris* 8 (1866).
3. Ibid.
4. Ibid.
5. Guillaume Bigourdan, "Notice sur la vie et travaux de M. Ch. Wolf," *Comptes rendus des séances de l'Académie des sciences* 157 (1918).
6. For the history of Wolf's machine and the debates over his work, see Jimena Canales, "Exit the Frog, Enter the Human: Physiology and Experimental Psychology in Nineteenth-Century Astronomy," *British Journal for the History of Science* 34 (June 2001), and Jimena Canales, "The Single Eye: Re-evaluating *Ancien Régime* Science," *History of Science* 39 (March 2001).

The event was exceptional because of its rarity: transits of Venus across the sun occur only approximately twice every hundred years. If astronomers missed their opportunity during the 1874 and 1882 transits, they would have to wait for the year 2004.

The desire to understand tenth-of-second and shorter moments of time intensified during this period.[7] The transit of Venus was the most important test for new techniques and methods designed to seize fugitive events and eliminate tenth-of-a-second differences in observations. The challenge consisted in timing the precise moment of the apparent contact between Venus and the sun—but tenth-of-a-second differences in determining this moment proved that the task at hand was particularly difficult. These same difficulties plagued a long list of other sometimes rare or sensational spectacles: eclipses, occultations, comets, and planetary phenomena such as sunspots, solar prominences, and flares—all "animated with movements of such violence that our terrestrial phenomena can only provide us with a weak comparison."[8]

The transit of Venus also promised to be scientists' best hope for solving the problem of standards since it was expected to close a century of debate surrounding the most important constant of celestial mechanics, the solar parallax.[9] A reliable figure for the solar parallax would enable

7. Given its importance at the time, the current historiography of the transits of Venus is surprisingly limited. Secondary sources include Albert Van Helden, "Measuring Solar Parallax: The Venus Transits of 1761 and 1769 and Their Nineteenth Century Sequels," in *Planetary Astronomy from the Renaissance to the Rise of Astrophysics, Part B: The Eighteenth and Nineteenth Centuries, the General History of Astronomy,* ed. René Taton and Curtis Wilson (Cambridge: Cambridge University Press, 1995); Peter Hingley and Françoise Launay, "Passages de Vénus, 1874 et 1882," in *Dans le champ des étoiles: Les Photographes et le ciel, 1850–2000* (Paris: Réunion des musées nationaux, 2000); Monique Sicard, "Passage de Vénus: Le Revolver photographique de Jules Janssen," *Études photographiques,* no. 4 (1998); A. Chapman, "The Transits of Venus," *Endeavour* 22, no. 4 (1998); David H. DeVorkin, "Venus 1882: Public, Parallax, and HNR," *Sky and Telescope* 22, no. 4 (1982), and *Henry Norris Russell: Dean of American Astronomers* (Princeton, N.J.: Princeton University Press, 2000), 44, 46, 49, 60; Steven J. Dick, Wayne Orchiston, and Tom Love, "Simon Newcomb, William Harkness, and the Nineteenth-Century American Transit of Venus Expeditions," *Journal for the History of Astronomy* 29 (August 1998); Paul M. Janiczek and L. Houchins, "Transits of Venus and the American Expedition of 1874," *Sky and Telescope* 48 (1974); Paul M. Janiczek, "Remarks on the Transit of Venus Expedition of 1874," in *Sky with Ocean Joined: Proceedings of the Sesquicentennial Symposia of the U.S. Naval Observatory, December 5 and 8,* ed. Steven J. Dick and LeRoy E. Dogget (Washington, D.C.: U.S. Naval Observatory, 1983).
8. Jules Janssen, "Les Méthodes en astronomie physique: Discours prononcé comme président du congrès le 26 août 1882," in *Lectures académiques: Discours* (Paris: Hachette et cie, 1903), 216.
9. "Parallax" generally refers to the angular change of an object when it is observed from two different positions. If the distance between the two observational positions is known, it can be considered the base of a triangle that when combined with measurements of the direction of the object as seen from both points can be used to determine the distance to the object. The solar parallax can be determined using Halley's method, which consisted in observing the transit of Venus across the sun. This was done either through the method of durations or the method of De l'Isle. In the method of durations the times of the transit as viewed from two different stations was determined, the lengths of the chords were deduced, and from these the least distance between the centers of the sun and Venus was found. However, since the method of duration required the observation of the whole transit, the method of De l'Isle was proposed. De l'Isle's

astronomers to determine the distance from the earth to the sun, set the dimensions of the solar system, and, using Newton's law, deduce the masses of the planets. Camille Flammarion, an astronomer and important popularizer of science, argued that with it astronomers would have "the meter of the *système du monde.*" The astronomer Hervé Faye agreed. The solar parallax was "the key to the architecture of the heavens" and an ultimate "touchstone, a precise verification of the theories of celestial mechanics."[10] At stake in this moment was nothing less than the determination of "the scale of the universe" and the problem of the plurality of worlds. Still more important, it was connected to philosophical debates about the value of geometric methods in astronomy and the nature of space and time—all lofty issues tied to earthly concerns of governance, national prestige, and military might.

Astronomers had long known that observing the transit of Venus was not the only way to establish a value for the solar parallax. But all other alternatives furnished radically different results. While observations of the previous transit (1769) moved most astronomers to settle on 8.57 seconds of arc, recent research on planetary movements as well as new determinations of the speed of light (1862) had led some to believe that the true value was around 8.86 seconds of arc.[11] In France, the Académie des sciences, the Bureau des longitudes, the Observatoire de Paris, and the École polytechnique all sponsored different types of evidence to determine the "true" value. To complicate matters further, proponents of these competing techniques

---

method required the precise determination of the time of contact between Venus and the sun at two different stations, which were either time coordinated or whose difference in longitude was accurately known. The precision of this method depended in great part on the accuracy of the longitude determinations.

10. Hervé Faye, "Association française pour l'avancement des sciences, congrès de Lille, conférences publiques: Le Prochain Passage de Vénus sur le Soleil," *Revue scientifique* 14, no. 16 (17 October 1874): 367–68. The problem of standards was articulated in Charles Wolf, "La Figure de la Terre, soirées scientifiques de la Sorbonne," *Revue scientifique* 7 (12 March 1870).

11. The Berlin astronomer Franz Encke assigned the value of 8.57 seconds of arc to the solar parallax from his analysis of the eighteenth-century transits, which ranged from 8.1 to 9.4 seconds of arc. Observations of Mars's oppositions against the sun convinced many astronomers to adopt a value exceeding 8.9 seconds of arc. Yet another method, using lunar theory, pointed to results around 8.916 (Hansen) or 8.850 (Stone) seconds of arc. Using planetary theory and the equation of the moon, Urbain Le Verrier, director of the Paris Observatory, set the solar parallax at 8.859 seconds of arc. By adopting this value he was able to reconcile discrepancies in the theories of Venus, Earth, and Mars. For descriptions of these methods, see George Forbes, *The Transit of Venus*, Nature Series (London: Macmillan, 1874); Edmond Dubois, "Nouvelle Méthode pour déterminer la parallaxe de Vénus sans attendre les passages de 1874 ou 1882," *Comptes rendus des séances de l'Académie des sciences* 69 (20 December 1869); Léon Foucault, "Détermination expérimentale de la vitesse de la lumière; Description des appareils," *Comptes rendus des séances de l'Académie des sciences* 55 (1862); Hervé Faye, "Note sur les nouvelles tables des planètes intérieures," *Comptes rendus des séances de l'Académie des sciences* 54 (1862); Urbain Le Verrier, "Sur les masses des planètes et la parallaxe du soleil," *Comptes rendus des séances de l'Académie des sciences* 75 (22 July 1872).

were often split intellectually as well as institutionally. Even when a group of scientists agreed on a certain method, differing observations within it rendered the results highly discordant. Who was right?

When astronomers during the late nineteenth century reviewed observations of the 1769 transit of Venus the conclusion was appalling: different people saw differently. Scientists, politicians, and even Napoleon III worriedly debated the nature of these differences, asking whether they were due to the fluctuating conditions of the phenomenon itself, to the different instruments employed, to the visual or mental apparatus of the observers, or, in some rare cases, to outright dishonesty.

Quite apart from the larger political and juridical consequences of disagreement, the immediate problems were insurmountable. If the solar parallax remained closer to the value determined by the Berlin astronomer Franz Encke in 1824, astronomers would have to posit the existence of an unlikely ninth planet. The problem was also directly connected to the long-standing riddle pertaining to Mercury's mysterious perihelion. According to Urbain Le Verrier, director of the Observatoire de Paris, only a "missing mass" could explain it. He believed that this mass might be found in the form of either a planet (commonly referred to as Vulcan) or of smaller intramercurial planets. He had hoped for an outcome that would match his previous triumph in the discovery of Neptune.[12]

Building on Le Verrier's work, Faye explained the looming crisis: "If we persist in our false evaluation of the parallax, the hypothesis that there exists some other planet which has gone unperceived until now . . . will need to be considered. And, since we cannot see such a probable planet, science finds itself forced into an impasse."[13] The astronomer argued that "all the discordances, all the contradictions which menace the future and, to some degree, the present of astronomy, will disappear" with an exact determination of the solar parallax.[14] Another possibility was equally dramatic. If the differences in the values for the parallax were caused by the Venusian atmosphere,

12. For histories of the problem of Mercury's perihelion that discuss Le Verrier's role see N. T. Roseveare, "Leverrier to Einstein: A Review of the Mercury Problem," *Vistas in Astronomy* 23 (1979); R. A. Lyttleton, "History of the Mass of Mercury," *Quarterly Journal of the Royal Astronomical Society* 21 (1980). On the so-called planet Vulcan and its implications for Newtonian theory, see William Sheehan and Richard Baum, *In Search of Planet Vulcan: The Ghost in Newton's Clockwork Universe* (New York: Plenum, 1997).
13. Faye, "Association française pour l'avancement des sciences, congrès de Lille, conférences publiques: Le Prochain Passage de Vénus sur le Soleil," 367. Faye also based his claim on recent work on the oppositions of Mars (George Biddell Airy), parallactic inequalities of the moon (Peter Andreas Hansen), and the speed of light (Léon Foucault).
14. Ibid.

then the possibility that Venus was a world like the earth would have to be seriously considered.[15]

As the century progressed, astronomers increasingly repudiated the mathematical methods that had characterized astronomy during the previous century. In the eighteenth century the British astronomer Edmond Halley had claimed that the solar parallax could be determined with exactness by combining simple Euclidean triangulations with direct observations of Venus's apparent contact with the sun. But more than a hundred years after Halley's discovery, astronomers came to doubt the possibility of timing the contact between two celestial bodies precisely. While some astronomers blamed the nervous systems of observers for discrepancies in results, others thought instruments were at fault. Still others believed that the problem was due to increasingly unskilled and undisciplined observers in astronomy. Some claimed that problematic nongeometric contacts arose from neither observational nor instrumental errors but were due to actual astronomical phenomena that needed investigation. Perhaps most alarmingly, some mathematicians and philosophers were led by the rift between the ostensibly straightforward geometric methods Halley had proposed and their apparently chaotic results to question the very foundations of mathematics. As geometric certainty became harder to obtain, many "solutions" were devised, including artificial transit machines for training an observer's responses or for measuring his delayed reactions and new cameras that photographed the event at short intervals. Yet none of these could eliminate insidious doubts as to the claims of scientific evidence to absolute truth.

Astronomers tried to solve reaction time and personal equation problems by developing, most intensively after the 1870s, new methods and techniques. In careful and costly preparations to eliminate tenth-of-a-second differences in observations, they adopted improved photographic techniques, hoping that with them they could finally determine precious astronomical constants.[16]

15. The astronomer and popularizer of science Camille Flammarion explained that if the results of the transits confirmed the existence of a Venusian atmosphere he would be led to the conclusion that "this planet is a world like ours." See Camille Flammarion, "Le Passage de Vénus: Résultats des expéditions françaises," *La Nature: Revue des sciences et de leurs applications aux arts et à l'industrie* 3 (1875): 394; Jules Janssen, "Sur l'éclipse totale du 22 décembre prochain," *Comptes rendus des séances de l'Académie des sciences* 71 (24 October 1870): 531, and "Note sur l'observation du passage de la planète Vénus sur le Soleil," *Comptes rendus des séances de l'Académie des sciences* 96 (29 January 1883).

16. For recent literature on photography in astronomy, see Jennifer Tucker, *Nature Exposed: Photography as Eyewitness in Victorian Science* (Baltimore: Johns Hopkins University Press, 2005); John Lankford, "The Impact of Photography on Astronomy," in *Astrophysics and Twentieth-Century Astronomy*, ed. Owen Gingerich, General History of Astronomy (Cambridge: Cambridge University Press, 1984); Holly Rothermel, "Images of the Sun: Warren De La Rue, George Biddell Airy, and Celestial Photography," *British Journal for the History of Science* 26 (1993); Alex Soojung-Kim Pang, "Victorian Observing Practices, Printing

Among the various methods proposed to eliminate individual differences in the appreciation of Venus's apparent contact with the sun, was a controversial new instrument, intriguingly named the "photographic revolver." Designed to photograph Venus's 1874 transit across the sun at intervals of approximately one second, the revolver was arguably the most promising device for ending the discord as to the exact time of the planet's apparent contact. It was invented by the astronomer Jules Janssen, known around the world as the famous scientist who left a besieged Paris in a hot-air balloon during the Commune uprising in order to observe an eclipse.[17] Widely considered the "missionary of the Bureau des longitudes" excelling in the study of "transient phenomena," Janssen was also well known for his pursuit of sunspots. In the 1870s he became deeply involved in a debate about the nature of sunspots and countered the claim advanced by some scientists, such as Le Verrier, that they were intramercurial planets.[18]

Technologies, and Representations of the Solar Corona, 1: The 1860's and 1870's," *Journal for the History of Astronomy* 25 (1994), and "Victorian Observing Practices, Printing Technologies, and Representations of the Solar Corona, 2: The Age of Photomechanical Reproduction," *Journal for the History of Astronomy* 26 (1995); Holmberg Gustav, "Mechanizing the Astronomer's Vision: On the Role of Photography in Swedish Astronomy, c. 1880–1914," *Annals of Science* 53 (1996); Alex Soojung-Kim Pang, " 'Stars Should Henceforth Register Themselves': Astrophotography at the Early Lick Observatory," *British Journal for the History of Science* 30 (1997), "Technology, Aesthetics, and the Development of Astrophotography at the Lick Observatory," in *Inscribing Science*, ed. Timothy Lenoir (Stanford, Calif.: Stanford University Press, 1998), and *Empire and the Sun: Victorian Solar Eclipse Expeditions* (Stanford, Calif.: Stanford University Press, 2002); Charlotte Bigg, "Photography and the Labour History of Astrometry: The Carte Du Ciel," in *The Role of Visual Representations in Astronomy*, ed. K. Hentschel and A. Wittman, Acta Historica Astronomiae (Thun: Deutsch, 2000). For a partial account of photography and the transits during this period that does not mention Janssen's photographic revolver, see John Lankford, "Photography and the Nineteenth-Century Transits of Venus," *Technology and Culture* 28, no. 3 (July 1987).

17. For Janssen, see Jimena Canales, "Photogenic Venus: The 'Cinematographic Turn' and Its Alternatives in Nineteenth-Century France," *Isis* 93 (2002); David Aubin, "Orchestrating Observatory, Laboratory, and Field: Jules Janssen, the Spectroscope, and Travel," *Nuncius* (2003); Françoise Launay, "Jules Janssen's In and Out Correspondence," in *100 Years of Observational Astronomy and Astrophysics: Homage to Miklos Konkoly Thege (1842–1916)*, ed. C. Sterken and J. B. Hearnshaw (Brussels: Sterken, 2001), *Un Globe-Trotter de la physique céleste: L'Astronome Jules Janssen* (Paris: Vuibert/Observatoire de Paris, 2008), and "Jules Janssen et la photographie," in *Dans le champ des étoiles: Les Photographes et le ciel, 1850–2000* (Paris: Réunion des musées nationaux, 2000); David Aubin, "La Métamorphose des éclipses de soleil," *La Recherche*, no. 321 (June 1999); Marie-Claude Mahias, "Le Soleil noir des nilgiri: L'Astronomie, la photographie et l'anthropologie physique en Inde du Sud," *Gradhiva: Revue d'histoire et d'archives de l'anthropologie* 24 (1998); Gérard Turpin, "Notes sur l'appareil de Janssen," *Les Informations de la Société française de photographie* 1 (1977); Françoise Launay, "Jules Janssen's 'Revolver Photographique' and Its British Derivative, 'the Janssen Slide,' " *Journal for the History of Astronomy* 36 (2005).

18. Victor Puiseux to ministre de l'Instruction publique, Paris, 12 February 1874, AN, F17 2928-2, Folder: M. Janssen mission dans l'Asie Orientale, éclipse de Soleil du 6 avril 1875, p. 2 ("missionary"); Hervé Faye, "Sur l'expédition de M. Janssen," *Comptes rendus des séances de l'Académie des sciences* 71 (12 December 1870): 821 ("transient phenomena"); Jules Janssen, "Passage de Vénus: Méthode pour obtenir photographiquement l'instant des contacts, avec les circonstances physiques qu'ils présentent," *Comptes rendus des séances de l'Académie des sciences* 76 (17 March 1873). For the debates about sunspots and intramercurial planets, see Urbain Le Verrier, "Bulletin des sociétés savantes, Académie des sciences de Paris.—2 octobre 1876: Les Planètes intra-mercurielles," *Revue scientifique* 18 (14 October 1876); Jules Janssen, "Bulletin des sociétés savantes, Académie des sciences de Paris.—2 octobre 1876: Observations sur les passages devant le Soleil de corps intra-mercuriels," *Revue scientifique* 18 (14 October 1876).

"No matter the cost"

Even before the problem of individual differences in observation leaked to the general public, governments across the world became concerned. Napoleon III's positivistic empire was the first in France to preoccupy itself with these strange divergences.[19] In 1869 the minister of public instruction, Victor Duruy, addressed a letter to the academy charging "scientific missionaries" to go to the end of the world in 1874 "to rid observations from the causes of error which so strangely affected those of 1769." Despite the "sorry state of the country's finances," the French government was able to amass an impressive amount of money and resources to overcome the obstacles that had haunted the observations made in the previous century. The problem, Faye explained, should be solved "no matter the cost."[20]

In 1866 the astronomer Charles Delaunay, an opponent of Le Verrier who would replace him as director of the Paris Observatory in 1870, inaugurated the debate in France with an article designed to point out the "embarrassment" of previous observations. According to Delaunay, a "black drop"

19. In 1866 a commission presided by the admiral Jurien de la Gravière was charged by the ministre de l'Instruction publique to aid astronomers in their observations of the next transit. On 15 July 1870 a *projet de loi* was sent to the Conseil d'État, which was interrupted by the war. Before the war the commission was composed of the admiral François Paris, Faye, Ernest Laugier, Antoine Yvon-Villarceau, and Puiseux, among others. See A. Gautier, "Travaux scientifiques étrangers: Travaux et faits astronomiques récents," *Revue scientifique* 8 (22 December 1871): 618; Victor Puiseux, "Note sur la détermination de la parallaxe du Soleil, par l'observation du passage de Vénus sur cet astre en 1874," *Comptes rendus des séances de l'Académie des sciences* 68 (8 February 1869). Puiseux's work was recounted in "Académie des sciences: Séance du lundi 8 février," *Les Mondes: Revue hebdomadaire des sciences et de leurs applications aux arts et à l'industrie par M. l'abbé Moigno* 19 (1869). See also *Journal officiel de la République française* (1 September 1872). "Projet de Loi" from Napoleon III, 14 June 1870, and Puiseux, "Rapport," 28 October 1866, Archives nationales (AN), F17 2928-1, Folder C: Commission de l'Académie des sciences, travaux, préparations, etc., pp. 1–10.

20. Ministère de l'Instruction publique, "Lettre," *Comptes rendus des séances de l'Académie des sciences* 68 (1 February 1869): 205. The meeting at the academy was recounted in "Académie des sciences: Séance du lundi 1er février," *Les Mondes: Revue hebdomadaire des sciences et de leurs applications aux arts et à l'industrie par M. l'abbé Moigno* 19 (1869). The minister's letter was reprinted in "Observation du passage de Vénus," *Les Mondes: Revue hebdomadaire des sciences et de leurs applications aux arts et à l'industrie par M. l'abbé Moigno* 19 (18 February 1869); Victor Duruy, *Journal officiel de la République française* (2 February 1869). The government initially gave 300,000 francs and in 1875 provided an additional 125,000 francs. The expenses, however, continued to mount, and in 1876 a *projet de loi* demanded 25,000 francs more from the Chambre des députés. For cost issues, see Faye, "Association française pour l'avancement des sciences, congrès de Lille, conférences publiques: Le Prochain Passage de Vénus sur le Soleil," *Revue scientifique* 14 (17 octobre 1874): 361; *Journal officiel de la République française*, no. 205 (27 July 1872; Chambre des députés, "Chambre des députés—annexe No. 445: Séance du 3 août 1876," *Journal officiel de la République française* (14 September 1876); "Loi ouvrant au ministre de l'Instruction publique et des Beaux-Arts sur l'exercice 1876 un crédit supplémentaire de 90,136 Fr. 39, applicable à des dépenses relatives à la détermination de la parallaxe du Soleil," *Journal officiel de la République française* 9, no. 85 (27 March 1877). The term "scientific missionaries" appeared in Jules Janssen, "Présentation de quelques spécimens de photographies solaires obtenues avec un appareil construit pour la mission du Japon," *Comptes rendus des séances de l'Académie des sciences* 78 (22 June 1874): 1731.

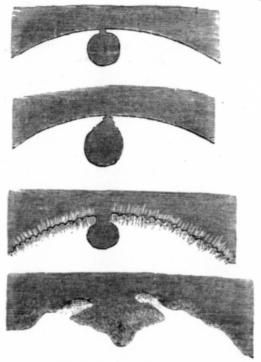

FIG. 15.—The "black drop," as observed in 1769.

Figure 4.1. The black drop seen during previous transits. From George Forbes, *The Transit of Venus*, Nature Series (London: Macmillan, 1874), p. 49.

that mysteriously appeared between Venus and the sun, combined with the problem of irradiation and personal errors in observations, contributed to the astronomers' "embarrassment in trying to determine the precise instant of contact" and caused an alarming "defectiveness of observations."[21]

In an article that appeared in the widely read journal *La Nature,* Wilfrid de Fonvielle, an important popularizer of science, described "that which was seen" in the transits of 1761 and 1769. He found discordant results even when astronomers observed side-by-side and with the same instruments. He suggested that the discrepancies were mainly due to the "black drop," which he described as "a mysterious object with very strange variations." Other astronomers similarly reported that "the transit of 1761 was

21. Charles Delaunay, "Notice sur la distance du Soleil à la Terre, extrait de l'Annuaire pour l'an 1866, publié par le Bureau des longitudes," in *Recueil de mémoires, rapports et documents relatifs à l'observation du passage de Vénus sur le Soleil* (Paris: Firmin Didot, 1874), 43, 97–98. The "goutte noire" was alternatively called the "ligament noir."

totally fettered" by the "black drop" phenomenon, yielding "all sorts of discordant results." Not only did observers disagree about what they saw, Faye complained, but "after a whole century of discussions, astronomers still have not been able to agree on the physical circumstances of the phenomenon, and on the true meaning of the important observations of 1769."[22]

The problem became particularly pertinent after a transit of Mercury visible in Europe (4 November 1868) produced discordant results. After presenting a brief history of the differences plaguing previous transit observations, Wolf categorically remarked: "Astronomy has not progressed since 1769." He blamed telescopic aberration and took a jab at his superior, Le Verrier, who for observing the transit had "placed himself in the same conditions as the astronomers of the previous century . . . saw the same phenomenon they saw, and obtained a number in accord with those of their observations."[23]

Discipline and Skill

To address the problem of divergent observations, the state-sponsored Transit of Venus Commission for determining the solar parallax, which was organized by the Académie des sciences, selected official methods and instruments. The Commission opted to rely chiefly on well-trained observers and on specific photographic instruments. In particular, it sponsored the work of Wolf, who had already addressed the problem of individual discrepancies in meridian transit observations.

In the face of a century of disagreements, in 1869 Wolf and his collaborator Charles André reported to the academy on their investigations to find "on which side truth lay." They concluded that observers timed this event differently primarily because their telescopes distorted the phenomena they were pointed at, not because of physiological differences. Wolf and André denied the existence of a physiological irradiation of the eye, which some had suggested could alter the estimation of time, and complained that it

22. Wilfrid de Fonvielle, "Les Derniers Passages de Vénus," *La Nature: Revue des sciences et de leurs applications aux arts et à l'industrie* 2, no. 43 (1874): 257, 59; Faye, "Association française pour l'avancement des sciences, congrès de Lille, conférences publiques: Le Prochain Passage de Vénus sur le Soleil," 365, and "Sur les passages de Vénus et la parallaxe du Soleil," *Comptes rendus des séances de l'Académie des sciences* 68 (4 January 1869): 42. Reprinted in Hervé Faye, "Académie des sciences: Séance du lundi 4 janvier," *Les Mondes: Revue hebdomadaire des sciences et de leurs applications aux arts et à l'industrie par M. l'abbé Moigno* 19 (1869).
23. Charles Wolf, "Le Passage de Vénus sur le Soleil en 1874," *Revue scientifique* 9 (20 April 1872): 1008. Eliminating differences in observations was seen not only as central to astronomy but as its defining measure of progress. Le Verrier observed the transit of Mercury from Marseille. He did not believe that telescopic aberration was the main problem, pointing out that astronomers had long differed in their measurements of the sun's diameter. He claimed the different measurements of diameters were in part responsible for the different times estimated during the transit. This theory was first stated in Urbain Le Verrier, "Théorie et tables du mouvement apparent du Soleil," *Annales de l'Observatoire impérial de Paris* 4 (1858): 69.

Observatoire de Paris

1843. V. Mauvais. — E. Bouvard	= — 0,18	
Id. — Goujon	= — 0,58	Arago, *Mémoire sur les erreurs personnelles,*
Id. — Laugier	= — 0,13	t. XI des *OEuvres*, p. 241
Id. — E. Liouville	= — 0,13	

V. Mauvais. — E. Bouvard	= + 0,05	
Id. — Goujon	= — 0,50	*Annales de l'Observatoire impérial,* t. II,
Id. — Laugier	= — 0,13	p. XLI.
Id. — Faye	= + 0,01	

1853. V. Mauvais. — Laugier	= — 0,31	
Id. — Goujon	= — 0,68	
Id. — Y. Villarceau	= — 0,20	*Annales de l'Observatoire impérial,* t. II,
Id. — Butillon	= — 0,22	p. XLI.
Id. — C. Mathieu	= — 0,59	
Id. — E. Liouville	= — 0,12	

1861. Le Verrier. — Y. Villarceau	= 0,00	
Id. — Chacornac	= + 0,11	
Id. — Lépissier	= + 0,26	*Annales de l'Observatoire impérial,* t. XVII
Id. — Besse-Bergier	= + 0,05	p. 20.
Id. — L. Folain	= 0,00	
Id. — Thirion	= + 0,11	
Id. — Ismaïl	= — 0,06	

1864. Le Verrier. — Y. Villarceau	= + 0,24	
Folain. — Y. Villarceau	= — 0,02	*Observations faites à deux instruments diffé-*
Folain. — Barbier	= — 0,05	*rents pour la détermination des longitudes.*
Lœwy. — Folain	= + 0,12	
Lœwy. — Wolf	= 0,00	

Figure 4.2. Personal equations at the Paris Observatory from 1843 to 1864. From Charles Wolf, "Recherches sur l'équation personnelle dans les observations de passages, sa détermination absolue, ses lois et son origine," *Annales de l'Observatoire impérial de Paris* 8 (1866): 153–208, on p. 184.

was "useless to give the name of a purely subjective phenomenon to a group of real phenomena linked to known causes." In this respect, they differed most noticeably from Faye, who had claimed that "the determinations [of astronomical phenomena] are complicated by a new personal error, varying from one observer to the next, and from one moment to the next for the same observer."[24]

24. Charles Wolf and Charles André, "Recherches sur les apparences singulières qui ont souvent accompagné l'observation des contacts de Mercure et de Vénus avec le bord du Soleil, Mémoire présenté à l'Académie des sciences, dans sa séance du 1er mars 1869," in *Recueil de mémoires, rapports et documents relatifs à l'observation du passage de Vénus sur le Soleil* (Paris: Firmin Didot, 1874), 125, 30. Their initial work was reported in "Académie des sciences: Séance du lundi 25 janvier," *Les Mondes: Revue hebdomadaire des sciences et de leurs applications aux arts et à l'industrie par M. l'abbé Moigno* 19 (1869). It was continued in "Académie des sciences: Séance du lundi 1er mars," *Les Mondes: Revue hebdomadaire des sciences et de leurs applications aux arts et à l'industrie par M. l'abbé Moigno* 19 (1869). And later recounted in Gautier, "Travaux scientifiques étrangers: Travaux et faits astronomiques récents," 618 n. 1, "Académie des sciences de Paris: Séance du 4 décembre 1871," *Revue scientifique* 8, no. 24 (9 December 1871); Charles Wolf, "Le Passage de Vénus sur le soleil en 1874, conférence faite à la Société des amis des sciences, le 29 mai 1873," in *Recueil de mémoires, rapports et documents relatifs à l'observation du passage de Vénus sur le Soleil* (Paris: Firmin Didot, 1874). A review appeared in Charles Wolf and Charles André, "Sur le passage de Mercure du 4 novembre 1868, et les conséquences à en déduire relativement à l'observation du prochain passage de Vénus," *Comptes rendus des séances de l'Académie des sciences* 68 (25 January 1869). The full report submitted to the academy appeared in *Comptes rendus des séances de l'Académie des sciences* 68 (1 March 1869). See also a letter describing the continuation of their work with artificial transits: André, 11 November 1876, Archives de l'Académie des sciences (hereafter cited as AAS), Carton 1645, Folder: Passage du Vénus, André.

Wolf and André claimed to have solved the "black drop" mystery with a new machine that artificially reproduced the transit of Venus across the sun that was modeled after the earlier prototype used to observe star transits. Like the previous one, this machine was also used to educate observers so as to reduce and stabilize their reaction times.

Wolf claimed that tenth-of-a-second delays arose from a combination of lack of training due to indiscipline and from the physiological phenomenon of visual persistence. He contested theories that claimed that these delays necessarily varied across time. Any variability, Wolf claimed, was due to the insufficient training of observers and could be eliminated through a strict disciplinary regime. His artificial star instrument was similar to one used by the Swiss astronomer Adolph Hirsch and that would later be adopted by experimental psychologists worldwide to measure reaction times. But he used it for a radically different purpose. While Hirsch used it to measure and correct for the personal equation, Wolf used it "for the education of young astronomers."[25] Hirsch, known for being a quite untalented observer, was less hopeful about the effect of education and entirely bewildered by Wolf's focus on discipline.[26] Why not, he asked, just measure and correct for the differences in observations, and "resign oneself to accept the sluggishness of the mind . . . as an unfortunate characteristic of every astronomer's nervous system."[27]

Wolf stressed the value of skilled and disciplined observation. In a historical work written toward the end of his life, the *Histoire de l'Observatoire de Paris,* which dealt with the period from the ancien régime to the French Revolution, he blamed "indiscipline" as a main cause behind a revolutionary delirium that had gripped the Observatory's assistants of the old regime and had plunged the Observatory into impotence until Le Verrier's rescue. In those revolutionary years, Wolf wrote, "indiscipline started to pen-

25. Wolf, "Recherches sur l'équation personnelle dans les observations de passages, sa détermination absolue, ses lois et son origine," 159.
26. Initially for Hirsch the temps physiologique was neither lower for skilled observers nor more regular. "Because even for *highly skilled observers* this time varies according to the observer's disposition at the time." Adolph Hirsch, "Expériences chronoscopiques sur la vitesse des différentes sensations et de la transmission nerveuse," *Bulletin de la Société des sciences naturelles de Neuchâtel (1861 à 1864)* 6 (1864): 103. Italics mine.
27. Adolph Hirsch, "Sur les erreurs personelles: lettre de M. A. Hirsch, directeur de l'Observatoire cantonal de Neuchâtel, à M. R. Radau," *Le Moniteur scientifique: Journal des sciences pures et appliquées* 8 (1866). Despite Hirsch's initial criticisms, Wolf's views on discipline were deeply influential. Hirsch's theory was later tempered by Wolf's work, and even Faye belatedly acknowledged how discipline—and not only mechanization—might solve the problem of individual differences in observation. Following Wolf, he admitted that the personal equation, after all, was nonexistent in the skilled musician or the disciplined soldier. Hervé Faye, *Cours d'astronomie et de géodésie de l'École polytechnique.* (Paris: Gauthier-Villars, 1883), 160.

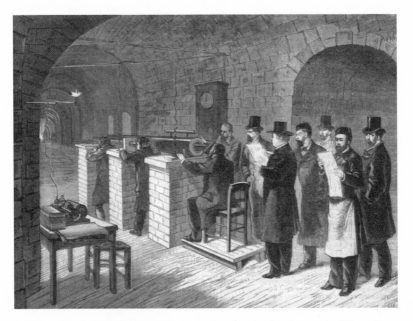

Figure 4.3. Personal equation tests in Paris in preparation for the 1874 transit of Venus. From *L'Illustration* 68 (1876), p. 392; reprinted in Simon Schaffer, "Astronomers Mark Time: Discipline and the Personal Equation," *Science in Context* 2 (1988): 115–45, on p. 125.

etrate the minds of students, incorporated to the national guard and lured into the revolutionary clubs."[28] After the Commune uprisings, indiscipline emerged as a national, historical problem linked to degeneration, ennui, and nervousness.

Wolf's artificial transit machine for time and longitude determinations was built to discipline observers and observations. The artificial transit of Venus machine was similar, but built on a larger scale.[29] Wolf aimed a telescope from the Paris Observatory at the Senate in the Jardin du Luxembourg (at a distance of 1,300 meters), where André operated a number of lamps and screens imitating Venus and the sun. When Wolf saw an "apparent" contact, he immediately pressed a telegraph key that sent the signal back to the Senate and compared it to the time of the "real" contact. From these experiments Wolf and André concluded that the "black drop" disappeared when an objective free of aberration was used (such as those made by Léon Foucault) and the instrument was aimed properly. Contradicting

28. Charles Wolf, *Histoire de l'Observatoire de Paris de sa fondation à 1793* (Paris: Gauthier-Villars, 1902), 328–29.
29. The Russians, Germans, Americans, and English also built artificial transit machines for this same purpose.

those who believed that the "black drop" was an inherent astronomical or physiological phenomenon, Wolf and André insisted that it was an illusion due mainly to defective telescopes and faulty aiming. They recommended that observers practice with moving targets and test for a personal equation in case it needed to be factored into the final result.

The most important conclusion of their paper was that the "black drop" was not a necessary impediment to observation and that, with the right instruments and training, observers could almost see the geometric contact expected, centuries earlier, by Halley when he claimed that the solar parallax could be unambiguously determined by observing Venus's apparent contact with the sun. For the moment, Wolf and André had vindicated the observational methods that were being so profoundly criticized by Faye and others. Furthermore, they concluded that there was no further point in investigating the physical aspects of the problem that fascinated Faye and the great popularizer of science Camille Flammarion. For Wolf, the existence of the "black drop" was nothing more than a "scientific prejudice." With deep irony he remarked, "The fable of an animal in the moon is still true."[30]

Wolf, who disagreed with Faye's interpretation of the personal equation, fought against Faye's description of a typical observer. He labeled Faye's theory of the personal equation as "psychological" and argued that errors of this nature were not omnipresent, as Faye claimed, but appeared only if the observer was not properly disciplined and skilled.[31] Instead of searching for quick technological fixes, Wolf argued for the importance of properly training observers. In his famous work on the personal equation he categorically stated that Faye's "psychological" theory of the personal equation and its supposed elimination by photography was flawed: "It seems impossible for me to continue to accept it as an explanation of the personal equation from which education does not free us."[32]

While Wolf denied that there was either an astronomical or physiological source for the "black drop," the issue of differences in observations persisted. In an article published in the popular *Revue scientifique,* he cautiously acknowledged that "this, nonetheless, is not to say that under

30. Wolf, "Le Passage de Vénus sur le Soleil en 1874," 1009.
31. The conceptualization of Faye's theory as "psychological" was continued by Gonnessiat and Sanford, who, by the turn of the century, described these astronomers as doing psychology. Ignoring essential differences between Faye's and Bessel's explanations, Sanford referred to them as "the same theory," which was, in essence, a "theory of the psychical cause of the personal equation." Edmund C. Sanford, "Personal Equation," *American Journal of Psychology* 2 (1889): 14, 413 n. 1; François Gonnessiat, *Recherches sur l'équation personnelle dans les observations astronomiques de passage,* Annales de l'université de Lyon (Paris: G. Masson, 1892), 3:119.
32. Wolf, "Recherches sur l'équation personnelle dans les observations de passages, sa détermination absolue, ses lois et son origine," 191.

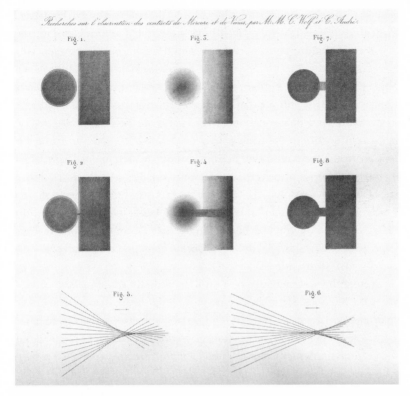

Figure 4.4. Illustration of Wolf and André's attempts to eliminate the black drop. From Charles Wolf and Charles André, "Recherches sur les apparences singulières qui ont souvent accompagné l'observation des contacts de Mercure et de Vénus avec le bord du Soleil, Mémoire présenté à l'Académie des sciences, dans sa séance du 1er mars 1869," in *Recueil de mémoires, rapports et documents relatifs à l'observation du passage de Vénus sur le Soleil*, 115–72 (Paris: Firmin Didot, 1874), p. 72.

these conditions observers will note exactly the same time, or experience the contact in the same way." In fact, his experiments showed that observers still did not time the contact identically, leading some scientists to the surprising conclusion that "the contact of two discs is never a purely geometrical phenomenon." Wolf argued that observers should be compared against each other "in order to determine their personal equations, if they exist." Le Verrier seconded his suggestion, arguing that astronomers should send on future expeditions "only those observers who have been compared among themselves."[33]

33. Wolf, "Le Passage de Vénus sur le Soleil en 1874," 1010; Urbain Le Verrier, "Sur les passages de Vénus et la parallaxe du Soleil," *Comptes rendus des séances de l'Académie des sciences* 68 (4 January 1869): 49.

According to one popular account of the attempts to observe the transit of Venus, Wolf and André's conclusions showed that "almost constant and VERY SIGNIFICANT differences persist between different observers, especially in estimating the time of . . . contacts." Even the experiments at the venerable Senate could not entirely eliminate individual differences. Rodolphe Radau, a scientist and popularizer who had earlier published influential essays on the personal equation and the speed of thought, discussed Wolf and André's work in the widely read *Revue des deux mondes:* "Nevertheless, there is a constant difference between the estimation of the moment of contacts by two observers—a difference due to physiological causes." In the end, the Commission was unable to muster full trust in Wolf and André's training machine, and it included photography as part of its effort to bypass the recalcitrant problem of individual differences in observation. Even Wolf and André, who did not advocate photography initially, became convinced of its usefulness.[34]

Despite the government's best efforts and hefty expenses, the controversy on methods and instruments pertaining to the "true" value of the solar parallax persisted well after 1874.[35] Even before the navy ships set sail to exotic places where the transit would be visible, optimism about the Commission's methods was fading.[36] One critic involved in the project who suspected that astronomers could not obtain the desired precision, demanded a "public discussion" in order to "eliminate all the doubts and all the errors."[37]

Against Geometry

The results of the transit opened up debates in mathematics and philosophy pertaining to the relationship of the mathematical and the physical

34. Wolf and André, "Sur le passage de Mercure du 4 novembre 1868, et les conséquences à en déduire relativement à l'observation du prochain passage de Vénus"; Edmond Dubois, *Les Passages de Vénus sur le disque solaire* (Paris: Gauthier-Villars, 1873), 154; Rodolphe Radau, "Le Passage de Vénus du 9 décembre 1874," *Revue des deux mondes* 1 (15 January 1874): 446. Reprinted in Rodolphe Radau, "Le Passage de Vénus du 9 décembre 1874," *Le Moniteur scientifique* 16 (1874).

35. In Britain George Biddell Airy at first determined the parallax to be 8.75 seconds of arc from the direct observations of the English expeditions, then settled on 8.82 seconds of arc. Simon Newcomb, in the United States, took as a mean value 8.85 seconds of arc; he then weighted this result against other methods and came up with a total value of 8.80 seconds of arc, which was adopted at the conférence internationale des étoiles fondamentales held in Paris (1896). See Steven J. Dick, Wayne Orchiston, and Tom Love, "Simon Newcomb, William Harkness, and the Nineteenth-Century American Transit of Venus Expeditions," *Journal for the History of Astronomy* 29 (August 1998): 248.

36. The places were Saint Paul and Campbell, Beijing (Peking), New Caledonia (Nouméa), Vietnam (Saigon), and Japan.

37. Montagne fils (*mécanicien*) to ministre de l'Instruction publique, St. Etienne, 29 January 1874, 26 February 1874, 9 April 1874, AAS, Carton 645, Folder: Passages de Vénus, demandes de missions.

world. Mathematics, according to Le Verrier, was a more powerful tool in astronomy than raw observation. It had proved invaluable during his discovery of Neptune, when he successfully predicted the place where the planet could be found in the sky. Almost immediate confirmation of its existence arrived with a sighting from Britain, demonstrating, to an awestruck public, astronomy's eerily predictive powers. Le Verrier maintained that the mathematical logic of celestial mechanics should be the ultimate arbiter in other debates, such as in controversies over sunspots, the perihelion of Mercury, and the solar parallax. The importance given to mathematics by the director of the Paris Observatory was shared by other scientists, most of who would have agreed with Flammarion's remark: "Geometry has justified its name by gaining possession of the terrestrial globe."[38] But in the case of the transit of Venus, traditional mathematical methods proved unsuccessful.

As mathematical methods were increasingly considered impotent at solving the problem, scientists questioned the presumed superiority of mathematics when compared to experimental astronomy. The famous doctor Claude Bernard was one of the first scientists to criticize astronomy for not being sufficiently experimental. All sciences, he explained, had to "pass by the stage of the sciences of observation," but only some could advance to the next, experimental level.[39] Bernard's *Principes de médecine expérimentale* was critical of astronomy.[40] While it was "expectant and passive," experimental science was "conquering and powerful."[41] Astronomers increasingly agreed. According to Faye, the problem of differing observations arose because "[astronomers] had reasoned too much like mathematicians."[42] Even Wolf, who came close to vindicating Halley's proposed "geometrical" methods with the objectives of Foucault, grew to accept that "the contact of two discs is never a purely geometrical phenomenon." And Victor Puiseux, the expert in mathematical astronomy who challenged the transit's photographic results, explained how, "in reality," the apparent contact of

38. Flammarion's defense of geometry was based on its importance in longitude determinations: Camille Flammarion, "Le Prochain Passage de Vénus et la mesure des distances inaccessibles," *La Nature: Revue des sciences et de leurs applications aux arts et à l'industrie* 2 (1874): 388.

39. Claude Bernard, *Principes de médecine expérimentale* (Paris: Presses Universitaires de France, 1947), 2.

40. Ibid.

41. Ibid.

42. Faye, "Association française pour l'avancement des sciences, congrès de Lille, conférences publiques: Le Prochain Passage de Vénus sur le Soleil," 365. The criticism of "mathematical conceptions" appeared earlier in Hervé Faye, "Sur les passages de Vénus et la parallaxe du Soleil," *Comptes rendus des séances de l'Académie des sciences* 68 (4 January 1869): 49. See also "Académie des sciences de Paris.—4 décembre 1871," *Revue scientifique* 8 (1871).

Venus with the sun does not occur with "the geometric simplicity that had been supposed."

Criticisms against Le Verrier, mathematical astronomer extraordinaire and widely disdained for his authoritarianism and antirepublicanism, translated rapidly into criticisms of mathematical forms of evidence. The popularizer of science Wilfrid de Fonvielle was most vocal, objecting that during the Second Empire "geometry has taken up arms" against astronomy. He demanded the elimination of the "radical subversion" of mathematics.[43] Delaunay, who briefly replaced Le Verrier, was also criticized for his antiobservational policies and mathematical leanings. Wolf privately communicated to the secrétaire général the faults of his superior Delaunay: "Recently, another example has presented itself. Although I have less liberty to speak of it, I will trust your utmost discretion and tell you the whole truth. M. Delaunay is even more of a stranger than his predecessor to the science of observation."[44] The *Revue scientifique* echoed these complaints. With the changes brought about by Delaunay, "physical astronomy has largely become the victim of the exuberant expansion given to mathematical astronomy."[45] The direction of astronomy changed in response to these ample criticisms, becoming increasingly experimental and further aligned with industrial and military concerns.

The Promise of Photography

During the decades immediately following the invention of photography, exposure times were regrettably slow. It took an average of two minutes to expose most daguerreotypes. Hence, most moving objects could not be captured. The exposure time of photography decreased gradually. It was not until the later part of the century that researchers started to see its potential not only in comparison to vision but, increasingly, in terms of its relation to temporal phenomena—including reaction time.

After Arago's death, Faye became the main advocate of scientific photography. Disillusioned by the transitory nature of the "fleeting instants of calm which the English astronomers call *a glimpse*," he promoted "the simple yet fecund idea of suppressing the observer and of replacing his eye

43. Fonvielle's criticism appeared in Wilfrid de Fonvielle, *L'Astronomie moderne* (Paris: Germer Baillière, 1868), xxxvi.

44. Charles Wolf to the secrétaire général, "Note sur les principes qui doivent régler le mode du gouvernement d'un observatoire," 3 March 1871, AN, F17–3720, Folder: Organisation 1872 (Décret du 5 mars), 2.

45. "Académie des sciences de Paris.—4 décembre 1871," *Revue scientifique* 1 (1871): 574.

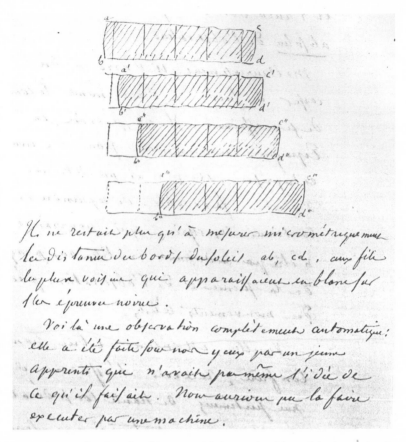

Figure 4.5. Letter from Faye describing sequential photographs of the sun's border crossing the wires of a telescope. From Hervé Faye, Paris, 31 October 1861, Preußische Staatsbibliothek zu Berlin, Sammlung Darmstädter I 1846(6) Faye, 3.

and brain with a sensitive plaque connected to an electrical telegraph."[46] In a set of pioneering experiments, he attached a photographic apparatus to a meridian telescope, and, by having an assistant press a key, automatically exposed the photographic film. Using an elaborate clockwork mechanism, he registered by telegraph the time of the "spontaneous" exposure. With "the simple movement of a handle" multiple photographs separated

46. Hervé Faye, "Sur l'observation photographique des passages de Vénus et sur un appareil de M. Laussedat," *Comptes rendus des séances de l'Académie des sciences* 70 (14 March 1870): 541. Italics in the original. Hervé Faye, "Rapport sur le role de la photographie dans l'observation du passage de Vénus, extrait des comptes rendus hebdomadaires des séances de l'Académie des sciences, séance du 2 septembre 1872," in *Recueil de mémoires, rapports et documents relatifs à l'observation du passage de Vénus sur le Soleil* (Paris: Firmin Didot frères, fils, et Cie, 1874), 228.

by one-second intervals were recorded in a single plate:[47] "Voilà, a completely automatic observation produced under our eyes by a young apprentice who had no idea what he was doing. We could have done it with a machine."[48]

Faye claimed that the best way of eliminating the "personal equation" was to "suppress the unfaithful eye of the observer" with photography where "everything is automatic." He "asked astronomers . . . to eliminate the human machine, whose imperfections are revealed to us in an alarming way."[49] Faye's description of a typical scientific observer included a fragile person, accosted by his unconscious, troubled by processes of "digestion, blood circulation or nervous fatigue," and whose "attention" was always limited. Faye's postlapsarian observer was naturally flawed and self-interested, seeing only what he wanted to see.[50] Photography compensated for an observer's attention deficit by permitting astronomers to consult records "at ease." His efforts to eliminate the effects of nervousness, distraction, excitement, and surprise led Faye to discard "the *ancienne* method based on our senses" and to instead advocate "automatic observation."

Photography should be employed by scientists precisely because it eliminated nervous feelings: "Here the nervous system of the astronomer is not in play; it is the sun itself that records its transit."[51] With photography, Faye claimed, "the observer does not intervene with his nervous agitations, anxieties, worries, his impatience, and the illusions of his senses and nervous system." Only by "completely suppressing the observer"—as photography purportedly did—could astronomers have access to nature itself: "[With photography] it is nature itself that appears under your eyes."

Defenses of photography use based on psychological explanations of observers' deficiencies gained prominence as the century advanced. The author of a widely read book on the transit of Venus expeditions explained

---

47. Hervé Faye, "Sur les passages de Vénus et la parallaxe du Soleil," *Comptes rendus des séances de l'Académie des sciences* 68 (11 January 1869): 71, 72. Reported in Hervé Faye, "Académie des sciences: Séance du lundi 11 janvier," *Les Mondes: Revue hebdomadaire des sciences et de leurs applications aux arts et à l'industrie par M. l'abbé Moigno* 19 (1869), and "Association française pour l'avancement des sciences, congrès de Lille, conférences publiques: Le Prochain Passage de Vénus sur le Soleil," 366. Other advocates of using photography for the transit of Venus were the British astronomer Warren de la Rue, the American Simon Newcomb, and the German Friedrich Paschen.

48. Hervé Faye, Paris, 31 October 1861, Preußische Staatsbibliothek zu Berlin, Sammlung Darmstädter I 1846(6) Faye, 3.

49. Hervé Faye, "Sur les erreurs d'origine physiologique," *Comptes rendus des séances de l'Académie des sciences* 59 (12 September 1864): 478.

50. "One only sees the things in which the observer is interested at the time of observation. The rest almost always escapes our unprepared attention." Ibid., 479.

51. Hervé Faye, "Sur les photographies de l'éclipse du 15 mars, présentée par M. Porro et Quinet," *Comptes rendus des séances de l'Académie des sciences* 46 (1858): 708.

how the drive toward exactitude produced anxiety and how photography was its palliative: "No nervous action can present itself to the photographer, comparable to the one that appears to the observer who observes or directly measures the position on Venus on the solar disk, and who finds himself under the influence of an anxiety which he feels when trying to give all the possible exactitude to his observation or measurements."[52] In contrast to an anxious astronomical observer, the astronomical photographer remained comparatively calm. Without this equipment, astronomers had to work under a "feverish hurry" since it was "difficult to keep one's *sang-froid* in the presence of the splendid phenomena that must be catalogued in a few minutes."[53]

The Transit of Venus Commission agreed with Faye.[54] The overwhelming conclusion of the Commission was that photographic observations would be better than other methods because they did not involve the observer's personal errors: "Photography," the astronomer Wolf concluded "is safe from this cause of error."[55]

Janssen, a friend and colleague of Faye who invented the "photographic revolver," agreed. Soon after its invention, Janssen's apparatus was modified and moved into other areas of science and culture, most famously to Étienne-Jules Marey's physiological laboratory and then to the studio of the Lumière brothers, where it was gradually transformed into what would soon be called the cinematographic camera.[56] From the moment Janssen pointed his revolver toward Venus (1874) to the time when he starred in one of the first films to be shown publicly (1895), the device passed through a painful gestation intimately tied to the debate on how to eliminate differing observations. The photographic revolver was an essential

---

52. Dubois, *Les Passages de Vénus sur le disque solaire*, 168.

53. Rodolphe Radau, "Les Applications scientifiques de la photographie," *Revue des deux mondes* 25 (15 February 1878): 882.

54. Among those who contributed to developing the photographic equipment used during the transit of Venus were J. G. Bourbouze, Cornu, Adolphe Martin, Fizeau, Wolf, and Yvon-Villarceau. Faye, "Association française pour l'avancement des sciences, congrès de Lille, conférences publiques: Le Prochain Passage de Vénus sur le Soleil," 366 ("observer does not intervene," "nature itself"), "Sur l'observation photographique des passages de Vénus et sur un appareil de M. Laussedat," 543 ("suppressing the observer"). For the commission's deliberations on photography, see Jean-Baptiste Dumas and Elie de Beaumont (Secrétaires perpétuels de l'Académie) to ministre de l'Instruction publique, Paris, 3 February 1873, AN, F17 2928-1, Folder C: Commission de l'Académie des sciences, travaux, Préparations, etc., p. 3.

55. Charles Wolf and Antoine Yvon-Villarceau, "Rapport sur les mesures micrométriques directes à faire pour l'observation du passage de Vénus, présenté à la Commission du passage de Vénus, dans la séance du 8 mars 1873," in *Recueil de mémoires, rapports et documents relatifs à l'observation du passage de Vénus sur le Soleil* (Paris: Firmin Didiot, 1874), 339.

56. Janssen's revolver is kept at the Conservatoire des arts et métiers. Revolver daguerreotypes of the transit of Venus are at the Paris Observatory. Photographs from and of the revolver can be found among the papers of Nadar at the Bibliothèque nationale de France.

part of a new, emerging, and highly contested evidentiary regime that, through chronophotography and its "inverse," sought to eliminate individual differences in the observation of moving phenomena by capturing tenth-of-a-second moments.

To Janssen's way of thinking there was something particularly photogenic about Venus. "For me," he explained, "it is the observation of the transit of Venus which specifically directed my attention to this fertile area." In a note published in the *Comptes rendus,* he explained his new procedure for observing the transit.[57] His method closely followed Faye's idea of photographing in a single plate sequences separated by a second.[58] Not only would this apparatus ostensibly suppress the observer, as Faye dreamed, but it also would permit the study of the physical circumstances surrounding the contacts. By aiming his revolver at artificial planets, Janssen "hoped that the photographic images [would] be free . . . of phenomena which so horribly complicate the optical observation of contacts."

The government emphatically sided with photography. By the mid-1870s the political tide turned decisively in its favor. Janssen, his revolver, and his calls for a physical astronomy based in large part on photographic methods became immensely popular—outside of the Académie des sciences. When Janssen returned from his expedition to Japan the president of the republic greeted him "warmly," and shortly thereafter the Meudon Observatory, with Janssen as director, was created by decree.[59] Following a plea from Faye on his behalf, Patrice de Mac-Mahon, president of the republic, gave him almost forty thousand francs to cover his outstanding expedition expenses and the instruments he had constructed for his personal use, including the revolver. Despite the Commission's criticisms of his methods, Janssen's photographic techniques eventually became powerful means for obtaining assent on visual matters, and the French government—reversing its earlier policy—

57. Janssen, "Présentation de quelques spécimens de photographies solaires obtenues avec un appareil construit pour la mission du Japon," 1730 (photogenic). The note was presented earlier to the Commission for the Transit of Venus; see Jules Janssen, "Méthode pour obtenir photographiquement les circonstances physiques des contacts avec les temps correspondants, communication faite à la Commission du passage de Vénus, dans sa séance du 15 février 1873," in *Recueil de mémoires, rapports et documents relatifs à l'observation du passage de Vénus sur le Soleil* (Paris: Firmin Didot, 1874).

58. At first Janssen tried a model built by Deschiens, which did not work properly since the instrument shook too much, and later he worked with the Rédier family to build the final prototype. For the transits of 1874 and 1882 Janssen also used a photographic telescope built by Steinheil. For the transit of 1882 Louis Pasteur, a photographer from the Meudon Observatory, aided him. This time he focused on the physical conditions of the phenomenon and not on determining the solar parallax.

59. Stanislas Meunier, "Académie des sciences," *La Nature: Revue des sciences et de leurs applications aux arts et à l'industrie* 3 (1875). Janssen's role in the transit of Venus expeditions was not the only motivation behind the creation of another observatory. Many factors came into play. The debate over the new observatory, for example, was even related to the question of transferring the Paris Observatory to a different location.

provided full support. By 1876 Janssen claimed that the revolver "was now definitely introduced in science."[60] A decade later he started to campaign for the Carte du ciel, a project for cataloguing stars with photography.[61] Ironically, this project was initially led by Ernest Mouchez, now director of the Paris Observatory, and one of photography's early critics. The debate pertaining to photography had been so fierce that Janssen explained its victory by reference to the wars of religion: "I will gladly say that you belong to . . . the triumphant church. But there was also amongst you a militant church, a church of catacombs, which the majority of you did not even know. And right now, your church triumphs as the Christian church has triumphed against Constantine."[62]

The defense of photography by the ministère de l'Instruction publique spread beyond astronomy. The government demanded that all its "scientific missionaries" adopt the new technology. By 1878 one observer remarked how scientists "today all carry a photographic apparatus." The ministry went so far as to create "a laboratory, in the ministry itself" to "initiate the travelers in the very simple manipulations of perfected processes." This governmental push was essential. Those with support from the government were being successful. But "the physicists and naturalists who work without this official backing are, in contrast, singularly embarrassed. In the presence of so many formulas and apparati whose number increases each day, they either hesitate to do photography or they make incorrect choices and are rapidly discouraged by their failures."[63]

The Cinematographic Turn

With his work on photography, Janssen wanted to do more than simply suppress the observer. He seized on one of the attributes of photography already advertised by François Arago in his famed speech to the Chambre des députés in 1839. Photography could expand the range of the visible. It could show a "new world" that corresponded much better to the "reality of things." The difference between ocular and photographic images was not a

60. Janssen to ministre de l'Instruction publique, 6 April 1876, AN, F17 2928–2, Folder: Commission chargée d'examiner les comptes des dépenses de M. Janssen, nommée le 16 mars 1876, p. 8.
61. The project of the Carte du ciel lasted until 1970, when the gargantuan effort of photographing the heavens was finally put to an end.
62. Jules Janssen, "En l'honneur de la photographie." Reprinted in Jules Janssen, "En l'honneur de la photographie: Discours prononcé au banquet annuel de la Société française de la photographie, juin 1888," in Oeuvres scientifiques recueillies et publiées par Henri Dehérain (Paris: Société d'éditions géographiques, maritimes et coloniales, 1930).
63. Ch. Fabre, Renseignements photographiques, Matériaux pour l'histoire primitive de l'homme (Paris: Ch. Reinwald et cie, 1878), 1.

problem, but a virtue, proving to him the "advantages of [the photographic] method."[64] Photography, according to Janssen, was "the true retina of the savant."[65]

Transforming photography into an evidentiary medium involved working on its reproducibility and standardization. While the Commission voted to use nonreproducible daguerreotypes for the 1874 transit, Janssen saw reproducibility as essential. In a speech given at the annual banquet of the Société française de photographie in 1888, he explained: "Of all [photography's] beautiful qualities, the most important perhaps is that of the conservation and multiplication of images."[66] The project of standardizing photography was part of Janssen's drive to eliminate personal—and national—differences in observations. He worked hard toward "unifying, simplifying, and standardizing" photography "as much as possible," in a project that included everything from instruments to terminology. In order to prove the advantages of photographic methods over observational astronomy, Janssen gave a lecture, later republished in the popular *Revue scientifique,* in which he showed how different astronomers had portrayed sunspots at different times. He started with a drawing by Fabricius (1611), moved successively through drawings by Galileo, Christoph Scheiner (1626), William Herschel (1801), and John Herschel (1837), and then after showing the divergent drawings of his contemporaries, concluded: "This short examination is sufficient for showing to us the disaccord that exists even amongst the best observers when observing solar phenomena. It convincingly demonstrates that the true method for observing them is to obtain, firstly, images drawn by the sun itself."[67] By shifting the "authorship" of the sun's images away from the scientists and to the sun itself, Janssen attempted to solve the problem of differing observations.[68]

64. Jules Janssen, "Sur la constitution de la surface solaire et sur la photographie envisagée comme moyen de découverte en astronomie physique," *Comptes rendus des séances de l'Académie des sciences* 85 (31 December 1877): 1250–51.

65. The phrase "true retina of the savant" was often repeated by Janssen and others; see, for example, J.-M. Eder, *La Photographie instantanée, son application aux arts et aux sciences,* trans. O. Campo, translated from the 2nd German edition. (Paris: Gauthier-Villars et fils, 1888), 221.

66. Janssen, "En l'honneur de la photographie." Janssen's defense of mechanically reproducible media increased over time. His revolver, for example, had worked with daguerreotypes. Since its initial conception, however, he was deeply concerned that his photographs be "comparable to those of other nations": Janssen, "Présentation de quelques spécimens de photographies solaires obtenues avec un appareil construit pour la mission du Japon," 1731.

67. Jules Janssen, *La Photographie céleste, conférence faite au congrès de Toulouse* (Paris: Administration de deux revues, 1888). Reprinted in Jules Janssen, "La Photographie céleste," *Revue scientifique* (14 January 1888).

68. For the question of scientific authorship, see Mario Biagioli and Peter Galison, eds., *Scientific Authorship: Credit and Intellectual Property in Science* (New York: Routledge, 2003).

Shooting at Stars

The debate pertaining to tenth-of-a-second differences in establishing moments of contact grew to encompass a larger debate about military prowess
and national superiority. In trying to capture events occurring within these
intervals, scientists faced the same difficulties as soldiers trying to hit targets or even march in synchronicity to drum beats. Transit observers were
compared to aiming snipers where a target had to be bisected with the sighting wires of a rifle in order to shoot at it.

Scientists compared telescopes to cannons, transit instruments to rifles,
and stars to moving targets. When studies on the personal equation revealed two different practices for observing star transits, one of them was
appropriately baptized the "shooting bird" method.[69] An astronomer writing in the interwar period explained that in most textbooks on descriptive
astronomy, "the flying bird and the sportsman, armed with a gun is the
popular illustration" and even referring to different species of birds could
be used to clarify astronomical practices: "Thus for a Mallard, dropping
into decoys, we allow a lead of a yard, while for the Black Brant, commonly
called a goose in California, we lead ten feet, because he is flying high and
his size makes his real speed very deceptive."[70]

The problem appeared urgent in terms of both offensive and defensive
strategies. How could a soldier react more quickly and resolutely to an enemy attack? Everyday, bodily reactions of a soldier trying to evade a bullet,
hit a target, or march in synchronicity to the beats of drum were related
to the tenth-of-a-second problems facing science. Astronomers sharpened
their aiming methods with artificial targets, while they suggested to the
armed forces ways of improving the techniques and instruments of the
art of war. Dodging bullets and trying to shoot at moving targets showed
how humans were sometimes one step behind reality.[71] Problems shared

69. Simon Newcomb classified the two methods: "In method A which I believe to be the most common
the observer anticipates the passage of a star, so as to have his signal completed at the exact moment
of crossing, much as if he were shooting at a bird. In the other method, which I shall call B, the observer
waits until he actually sees the star on the thread, and then commences the nisus which terminates in the
signal." In the flying bird method, the observer could accidentally signal before the star actually crossed
the wire, and consequently the personal equation would be smaller whereas the actual error in the determination of the star's position might have increased. Cited in Raynor L. Duncombe, "Personal Equation
in Astronomy," *Popular Astronomy* 53 (1945): 9.

70. R. H. Tucker and Knight Dunlap, "The Personal Equation and Reaction Times," *Science* 57, no. 1480
(1923): 557.

71. John Locke wrote about perceptual delays using the example of the imperceptibleness of a speeding
cannonball. "Let a cannon bullet pass through a room, and in its wake take with it any limb, or fleshy parts
of man; 'tis clear as any demonstration can be, that it must strike successively the two sides of the room;

by the military and science fundamentally changed not only the choice of problems in astronomy but also the types of evidence worthy for establishing scientific proof. The right astronomical methods and valuable scientific skills showed results on observatories as much as on battlefields.

Since midcentury the German physiologist Helmholtz and his colleague Emil du Bois-Reymond used the Melvillian image of a wounded whale to illustrate the dangers of reacting slowly. In whale hunting, the nerve impulse from a harpoon going into the animal's tail took about a "whole second" to reach the "brain of the monster."[72] As part of the new demands on science imposed by defeat, France reevaluated the relationship of science to the art of war. Earthly and astronomical concerns became purposefully intertwined, especially since French scientists—and French maps—had been implicated in defeat.

Tenth-of-a-second problems intensified after the Franco-Prussian War, when the problem of differing reactions in science was widely publicized and the educated public learned—in detail—of its dangers. The astronomer Faye chastised the military for lagging behind astronomy in its adoption of new optical technology. Shortly after the fall of the Commune, he returned to Paris to present a paper titled "On the art of targeting and its physiological characteristics" in which he stated innovations in astronomy that could be used for improving artillery and long-range missiles.[73] Aiming at stars and firing at distant targets encountered similar technical, disciplinary, and physiological problems. Astronomers, Faye explained, could aim at a star within .25 seconds of arc. If these scientific innovations were adopted by the military, targets at a distance of one kilometer could be closed in to within a single millimeter. Additionally, instead of the traditional "Ready! Aim! Fire!" he suggested that these three commands be broken into the five steps used in astronomy when determining time and longitude.

'tis also evident, that it must touch one part of the flesh first, and another after; and so in succession: and yet believe nobody, who ever felt the pain of such a shot, or heard the blow against the two distant walls, could perceive any succession, either in pain or sound of so swift a stroke." John Locke, *An Essay Concerning Human Understanding* (London: Penguin Books, 1997), 177.

72. Emil du Bois-Reymond, "Vitesse de la transmission de la volonté et de la sensation à travers les nerfs: Conférence de M. du Bois-Reymond à l'Institution royale de la Grande-Bretagne," *Revue scientifique* 4, no. 3 (15 December 1866): 39. For Helmholtz's use of the whale imagery, see Hermann von Helmholtz, "Ueber die Methoden, kleinste Zeittheile zu messen, und ihre Anwendung für physiologische Zwecke," *Königsberger Naturwissenschaftliche Unterhaltungen* 2 (1851): 325, and "On the Methods of Measuring Very Small Portions of Time, and Their Application to Physiological Purposes," *The London, Edinburgh and Dublin Philosophical Magazine and Journal of Science* 4 (1853): 189.

73. Hervé Faye, "Sur l'art de pointer et ses conditions physiologiques," *Comptes rendus des séances de l'Académie des sciences* 71 (19 December 1870).

Even the relation of reaction time to rhythm was thought of in both military and scientific terms. The astronomer who could time a star across the crosshairs of his telescope was described as a musician who met the sound of the director's baton, and both were compared to the disciplined, skilled soldier who could meet his target. François Gonnessiat, an assistant astronomer from Lyon who wrote a book on the personal equation, compared a talented astronomical observer to a musician and used military terms to describe both: "it is, in effect, extremely rare that the first *attack* of a [musical] piece be perfectly frank."[74] Henri Beaunis, the well-known experimental psychologist who was a *chef d'ambulance* during the war and lived close to the contested Franco-German frontier, reminded his students that "today, everyone is a soldier."[75]

The problem of determining inaccessible distances in astronomy, such as deriving the distance from the earth to the sun through observations of the transit of Venus, was militarily pertinent. New artillery proved the need for determining the distance of barely visible and inaccessible targets, and the efforts placed on transit of Venus observations were seen as a subset of this great and mundane concern.[76] In his public lectures on the transit Faye explained how astronomers acted like "artillerymen who need to determine the distance of the target if they want to aim in a way to hit it" and compared the telescope to a "cannon."[77] The astronomer Charles Wolf also explained the transit in the same military terms. "You know," he wrote, "how in the battlefield an officer deduces the distance of the enemy battalion from the angle at which he sees a man, whose average height he knows. This angle is, for the enemy battalion, the parallax of the officer."[78] The sentiment of extending scientific and military cooperation resulted in a profound collaboration between the navy and the Académie during the 1874 transit of Venus expeditions. These concerns continued for decades. The astronomer Janssen, for example, worried about determining inaccessible distances

74. Gonnessiat, *Recherches sur l'équation personnelle dans les observations astronomiques de passage*, 143. Italics original.

75. Henri Beaunis, *L'École du service de santé militaire de Strasbourg et la Faculté de médecine de Strasbourg de 1856–1870* (Nancy: Berger-Levrault et cie, 1888), 18.

76. Flammarion, "Le Prochain Passage de Vénus et la mesure des distances inaccessibles"; Radau, "Le Passage de Vénus du 9 décembre 1874," 435. Reprinted in Radau, "Le Passage de Vénus du 9 décembre 1874."

77. Faye, "Association française pour l'avancement des sciences, congrès de Lille, conférences publiques: Le Prochain Passage de Vénus sur le Soleil," 363.

78. Charles Wolf, "Conférence sur les applications de la photographie à l'astronomie et en particulier à l'observation du passage de Vénus," *Bulletin de la Société française de photographie* 21 (8 January 1875).

after the Boer war: "New infantry and especially artillery weapons" could reach "the enemy in places where he is sometimes barely visible."[79]

The competition to establish the precise moment of contact during the transit of Venus was described variously as a "lutte pacifique,"[80] a "European debacle,"[81] and "a gigantic scientific tournament."[82] While in the 1860s astronomers suggested that different nations work together, the Franco-Prussian War completely eliminated any hopes for international cooperation.[83] The fight for eliminating individual differences was a belligerent and nationalistic competition. Faye explained how for the coming transit British astronomers would base their observations on Halley's method of contacts, developed by "one of their most renowned compatriots." Germans, he continued, selected heliometers, "consecrated . . . by the memory of Bessel's beautiful measurements" and built by the famous optician Joseph Fraunhofer. The French choice, he prescribed, should be similarly patriotic: "I believe our approach should be the integral application of methods originally indebted to the discoveries of Daguerre, Arago, and Ampère. We will see to which nation will belong the honor of best serving science in this generous fight."[84] He explained why the "great governments of the worlds were interested" in this event: "So, the fate of [astronomy] and its progress are intimately connected to the naval, military and commercial power of each maritime state."[85] Alerted to the military, political, and scientific meaning of the problem, the French government did everything in its power to solve it.

Yet timing the moment of contact between two celestial bodies proved so difficult that some scientists, including the Belgian psychophysicist Joseph Delboeuf, were led to believe in an essential discontinuity in percep-

79. Jules Janssen, "De la nécessité actuelle de la topographie: L'Exemple des Boers," in *Lectures académiques: Discours* (Paris: Hachette et Cie, 1903), 295. This volume includes many other speeches related to war and the military. See, for example, "Assemblée générale des dames françaises pour les secours aux blessés," pp. 277–84.

80. Although the warring language was present in texts regarding the transit of Venus even before the Franco-Prussian War, afterward it intensified. Ministère de l'Instruction publique, "Lettre," 206; Wolf, "Le passage de Vénus sur le Soleil en 1874," 1011, and "Le Passage de Vénus sur le Soleil en 1874, Conférence faite à la Société des amis des sciences, Le 29 mai 1873," 401.

81. Faye, "Sur l'observation photographique des passages de Vénus et sur un appareil de M. Laussedat," 524.

82. "Académie des sciences de Paris: Séance du 29 juin 1874," *Revue scientifique* 14, no. 1 (4 July 1874): 23.

83. Ministère de l'Instruction publique, "Lettre," 206.

84. Faye, "Sur l'observation photographique des passages de Vénus et sur un appareil de M. Laussedat," 544. Others followed Faye in naming the methods English (Halley's), German (heliometry), and French (photographic). Dubois, *Les Passages de Vénus sur le disque solaire*, 161–62. Other methods included spectroscopic observations and the use of a double-image micrometer. A micrometer is a telescope accessory for measuring small angles.

85. Faye, "Association française pour l'avancement des sciences, congrès de Lille, conférences publiques: Le Prochain Passage de Vénus sur le Soleil," 366–67.

tions. This problem exacerbated after the transit of Venus: "One knows that, in the observations of the transit of Venus over the sun, astronomers are frustrated by a phenomenon know as the black drop." Like two drops of water that "once they are sufficiently close to one another get brusquely confused," Venus's apparent contact with the sun happened *as* suddenly. Delboeuf explained how Venus seemed suspended when it was at the verge of an apparent contact, and then "tout à coup" it seemed to "precipitate" toward the sun, "in a manner which made it impossible to know the precise instant."[86] While Delboeuf concluded that perceptions were "necessarily discontinuous," numerous other scientists started searching for new technologies to overcome these perceptual deficiencies.

The astronomer Adolph Hirsch complained that "each nation had come up with their own solar parallax." And the popular science writer Wilfrid de Fonvielle, who had alerted the general public to the discordances of the previous transits of Venus, mocked the astronomers' hopes. In *Le Mètre international définitif* he commented cynically on the host of solar parallax values that had resulted from the British, American, French, German, and Russian expeditions: "There are as many great nations as there are distances from the sun to the earth. It is terribly irritating that each nation cannot have its own special planet for its own individual use and is obliged to prosaically receive heat from that banal celestial body which illuminates all the others."[87] The attempt to determine an absolute standard of measurement from observations of the transit of Venus suffered a fate similar to the earlier attempt to deduce the meter from measurements of the circumference of the earth, forcing scientists to reevaluate their claims to absolute truth. The tenth of a second remained as mysterious as ever.

86. Joseph Delboeuf, "La Loi psychophysique et le nouveau livre de Fechner," *Revue philosophique de la France et de l'étranger, dirigée par Th. Ribot* 5 (1878): 134.
87. Wilfrid de Fonvielle, *Le Mètre international définitif* (Paris: Masson, 1875), 140. Adolph Hirsch, Ministère de l'Instruction publique, "Première séance: Mercredi 5 octobre 1881," in *Conférence internationale du passage de Vénus* (Paris: Imprimerie nationale, 1881), 10.

CHAPTER 5

CAPTURED BY CINEMATOGRAPHY

When a horse passes in front of us in a gallop, the forms that it assumes
every *tenth of a second* fuse in our eye into a unique image.

LOUIS OLIVIER, director of the Revue générale des sciences pures et appliquées

By the end of the century, new techniques permitted scientists to photograph high-speed objects—increasingly at intervals briefer than a tenth of
a second. In 1851 Henry Fox Talbot had used an electric spark as a flash,
photographing a page of the *Times* while it was pinned to a rapidly rotating wheel. To general amazement, he was able to read the newspaper's text
in the photograph. Similar methods using sparks as photographic flashes
were employed later in the century by Ernst Mach, Charles Vernon Boys,
Lord Rayleigh, and Arthur Worthington.[1] Other techniques involved using
extraordinary powerful sources of illumination, such as the sun. By 1876
Janssen's photographs of the sun were already obtained in elapsed times

1. For the photographic work of Charles Vernon Boys, Lord Rayleigh, and Arthur Worthington, see Simon
Schaffer, "A Science Whose Business Is Bursting: Soap Bubbles as Commodities in Classical Physics," in
Things That Talk: Object Lessons from Art and Science, ed. Lorraine Daston (New York: Zone Books, 2004).
For the speed and schlieren photographs of Mach, see Christoph Hoffmann, "Mach-Werke," *Fotoges-
chichte* 60 (1996): 3–18. Since a spark's field of illumination could only be extended by a few inches, the
repertoire of these scientists remained quite limited, mostly to splashes and bullets.

of less than one hundredth of a second.[2] Others used multiple cameras or lenses. In this way Eadweard Muybridge, working in California, produced the famous photographs of running horses that captivated the world. Albert Londe, working at the Salpêtrière Hospital, employed multiple lenses to capture fleeting moments. In Germany, Ottomar Anschütz used a similar technique to photograph animals in motion. In 1882, Étienne-Jules Marey, fascinated by Muybridge's images, transformed Janssen's revolver into a faster *fusil photographique*. The speed of photography and cinematography continued to increase: in the 1930s strobe lamps were used to capture intervals smaller than a thousandth of a second.

Instruments designed to visualize tenth-of-a-second and much shorter periods of time solved certain problems, yet they introduced others. Gaston Bachelard, a controversial scientist, philosopher, and poet writing during the stroboscopic era, described the task at hand as requiring thinkers to question "the ease . . . of correspondence between 'real' phenomenon and the instrumental phenomenon of stroboscopy."[3] For a number of scientists who created some of these instruments and who evaluated their merits, particularly the benefits and disadvantages of photography and cinematography, their work involved thinking about the tenth of a second.

Although late nineteenth-century high-speed photography frequently surpassed the tenth-of-a-second threshold, the use of photography and cinematography in science was regularly justified by reference to this interval. Photography was superior to direct observation, some scientists argued, because of the eye's inability to store light for intervals longer than a tenth of a second and because of its inability to see events taking place at intervals shorter than this value.

The assessment of Arago, who introduced daguerreotypes to the general public, was typical. He noted that one essential property of the eye was its inability to register moments shorter that the tenth of a second: "The human eye . . . is constituted in such a way that a luminous sensation does not disappear until *a tenth of a second* has passed after the complete elimination of the cause that produces it."[4] And in the last book on which he worked, he carefully analyzed the well-known "children's game, that is

2. Davanne, "Rapport sur la XIe exposition de la Société française de photographie," *Bulletin de la Société française de photographie* 22 (1876): 227.
3. Gaston Bachelard, *The Dialectic of Duration*, trans. Mary McAllester Jones (Manchester: Clinamen Press, 2000 [1936]), 78.
4. François Arago, "Notice scientifique sur le tonnerre," *Annuaire du Bureau des longitudes* (1838). Cited in Léon Lalanne, "Note sur la durée de la sensation tactile," *Journal de l'anatomie et de la physiologie normales et pathologiques de l'homme et des animaux* 12 (1876): 453. Italics mine.

to say, the experiment which all have done or have seen done, and which consists in producing a continuous ribbon of light by the rapid movement of a glowing ember."[5]

According to the astronomer Jules Janssen, photography was infinitely superior to the eye because the latter was inescapably limited by the tenth of a second. While "all impressions that are dated by a tenth of a second are effaced," photography could store them for much longer. The eye sensed light only for "approximately a tenth of a second, since effects accumulate on the retina after the beginning of the luminous impression until this time lapses." Photographs, in contrast, could capture more light by being exposed for longer periods.[6] Agnes M. Clerke, a famous American popularizer of astronomy and champion of astronomical photography, expressed a similar idea. She criticized how the eye's visual persistence was limited by "*one-tenth part of a second,* leaving it a continually renewed tabula rasa."[7] Photography, in contrast, had much higher retentive and storage powers.

To demonstrate that the tenth of a second was so integral to vision, Janssen speculated about how our experience of the world would be altered by a change in this value. "The alarming consequences of a simple change in the duration of retinal impressions" could be disastrous for most humans: "the light of day would become unbearable." But it could be beneficial for night watchers: "the night would show constellations of stars. Our celestial vault would appear like an immense milky way."[8]

Cinematographic research was also described by reference to the tenth of a second. By recording and displaying images at speeds close to this value, researchers argued, these technologies captured movement. The physiologist Marey explained how "because our retina is incapable of perceiving clearly more than ten images per second" new instruments could provide a viewer with the illusion of movement if they were passed before the viewer's eyes at this speed.[9] To produce the illusion of movement, he explained,

5. François Arago, "Le Tonnerre," in *Oeuvres complètes* (Paris: Gide et J. Baudry, 1854), 59. For the context of this work see François Arago, "Avertissement des éditeurs," in *Oeuvres complètes* (Paris: Gide et J. Baudry, 1854), iv.

6. Jules Janssen, "Les Méthodes en astronomie physique: Discours prononcé comme président du congrès le 26 août 1882," in *Lectures académiques: Discours* (Paris: Hachette et cie, 1903), 211–21, on pp. 217–18.

7. Agnes Clerke, *A Popular History of Astronomy during the Nineteenth Century* (Edinburgh: Adam and Charles Black, 1885), 451. Italics mine. Cited in Bernard Lightman, "The Visual Theology of Victorian Popularizers of Science: From Reverent Eye to Chemical Retina," *Isis* 91 (2000): 677.

8. Jules Janssen, "Les Méthodes en astronomie physique: Discours prononcé comme président du congrès le 26 août 1882," in *Lectures académiques: Discours* (Paris: Hachette et cie, 1903), 218.

9. Étienne-Jules Marey, "La Photochronographie et ses applications à l'analyse des phénomènes physiologiques," *Archives de physiologie normale et pathologique, publiées par MM. Brown-Séquard, Charcot, Vulpian* 1 (1889).

"at least ten or twelve images per second" needed to be displayed.[10] The unraveling of sound and speech across time, one scientist surmised, functioned analogously to that of images—both could be imitated by stringing together tenth-of-a-second snapshots. A psychophysicist who worked at the Société de psychologie physiologique explained how the possibility of combining moving images with voice was not only possible but imminent. He expected both images and sound to be recorded and played back at intervals "equal to the tenth of a second": "because there is not a syllable which requires less than a tenth of a second to be pronounced."[11]

The tenth of a second entered into a much larger debate about the objectivity of photography and cinematography and their place within art and science. A prominent editor of a scientific journal explained the distinction between art and science by reference to this value. He concluded that, because of the tenth-of-a-second limitation of vision, there would always be two distinct truths. He termed one of them scientific, objective, and photographic, and the other one artistic and subjective. "When a horse passes in front of us in a gallop," he explained "the forms that it assumes *every tenth of a second* fuse in our eye into a unique image." Direct, subjective vision could not peer beyond this tenth-of-a-second limit. This limitation was so essential that it marked the difference between scientific and artistic forms of representation. Artistic depictions of movement should display these tenth-of-a-second limitations, while scientific ones should surpass them: "In effect, paintings should represent objects as we see them, that is, if the movement is fast, as they are not. . . . If this somewhat confused image is faithfully rendered by the painter, it will provide us with the sensation which we had in seeing."[12]

Defenses of photography and cinematography based on the tenth of a second's relation to vision were part of a larger effort to establish their trustworthiness. This effort involved many fronts, such as standardization and publication practices, and examining the role of authorship, memory, and experience.

Scientists such as Janssen and Marey succeeded in establishing guidelines for the use of photography and cinematography that would be widely followed. Janssen, for example, standardized photographic methods, con-

10. Étienne-Jules Marey, "Fusil photographique," *Bulletin de la Société française de photographie* 28 (1882).

11. Georges Guéroult, "Sur une application nouvelle de la photographie et du phénakistiscope," *Comptes rendus des séances de l'Académie des sciences* 122 (1896), and "Sur un moyen d'emmagasiner les gestes et les jeux de physionomie," *Comptes rendus des séances de l'Académie des sciences* 108 (1889).

12. Louis Olivier was director of the *Revue générale des sciences pures et appliquées*. Louis Olivier, "Physiologie: La Photographie du mouvement," *Revue scientifique* 30 (1882): 809. Italics mine.

strued scientific images as authorless, and separated them as much as possible from artistic practices. Marey, in turn, distinguished scientific and
unscientific uses of photography by claiming that the former should be
unedited, should record its objects from a single point of view, should
have equal intervals between frames, and by arguing that a single machine
should be used for both projecting and recording.

These guidelines were frequently contested by lesser known scientists who pioneered alternative imaging techniques. For example, Étienne
Léopold Trouvelot, who would engage Janssen in a bitter struggle over scientific images, held a different point of view with regards to many of these
elements. Commercial cinematographers soon alienated Marey by introducing their own changes to cinematographic technologies. After separating
the recording and projecting functions of the machine, they also turned to
Marey's bête noire, editing.

The Larger Problem

Behind the history of the technological success of photographic imaging
techniques lies a different story, which is mostly still hidden in archives,
revealing that—within closed rooms—astronomers acknowledged that
their main weapon, photography, had failed to completely demystify this
moment. Different cameras produced different results. Which should be
trusted? Should photography be standardized, or should it remain a mallea-
ble craft? Should scientists publicize all photographs or only a few selected
ones? Should they reveal disagreements among themselves or portray a
cleaner view of the scientific method to the public?

The 1874 transit of Venus was an important trial for astronomical photography, for which astronomers devised new instruments and techniques
and improved on preexisting ones. Yet defending the importance of photography for the exact sciences became particularly hard after photographs
from the transit showed mostly discordant results. Despite the government's best efforts, inconclusive results from the expeditions proved to
even the most credulous the difficulties—perhaps even the impossibility—
of defining tenth-of-a-second moments of time and of finding an absolute
standard of measurement. To add urgency to the matter, the next transit
of Venus (1882) was rapidly approaching—the last until the twenty-first
century.

The hopes of photography's advocates proved to be premature. The
worst fears of its defenders were realized when it became evident after
the transit that the different cameras used had produced photographs so

different that it was impossible to compare their results.[13] The same singularities that had plagued the transit of Venus in the previous century reappeared, and scientists were unable to determine the "real" instant of the planet's apparent contact with the sun.

Astronomers did not give up: a new commission was created to deduce the parallax from measurements taken from photographs—again with alarmingly discouraging results.[14] The physicist Armand Fizeau, who was in charge of the project, explained that it was going "slowly, but surely." But Victor Puiseux, a mathematician who had worked at the Bureau des calculs of the Bureau des longitudes, challenged the photographic results, considering them inferior to those obtained from direct observations.[15] In the end, photography's advocates gave up, announcing that the feared personal equation reappeared in efforts to measure photographs. They now concluded that the problem had been merely shifted from discrepancies in the direct observations to those in measurements taken from photographs. Urbain Le Verrier, director of the Paris Observatory, and the physicists Fizeau and Alfred Cornu increasingly distanced themselves from photo-

13. Warren de la Rue had unsuccessfully advocated that astronomers use similar instruments in order to have comparable photographs. George Biddell Airy wanted the telescopes used in the British expeditions to be as similar as possible. For some attempts toward standardization, see Edmond Dubois, *Les Passages de Vénus sur le disque solaire* (Paris: Gauthier-Villars, 1873), 156.

14. *Documents relatifs aux mesures des épreuves photographiques*, 3 vols., vol. 3, Recueil de mémoires, rapports et documents relatifs à l'observation du passage de Vénus sur le Soleil, extrait du tome III, 3e partie (Paris: Gauthier-Villars, 1882). Experiments on taking readings from photographs had been carried out in preparation for the transit, where Charles Wolf and Antoine Yvon-Villarceau concluded that photographs should be measured by the same observer. The sheer number of photographic plates, however, later required that more than one observer be employed. Similarly, de la Rue, Balfour Stewart, and Maurice Loewy all carried out investigations on the deformation of photographic plates. Some critics noticed how the exposed photographic plates suffered from some of the same deficiencies as eyes—like irradiation—and similarly concluded that the personal equation of observers reappeared when measurements were taken off photographs. In 1876 Alfred Angot warned the Commission of the Transit of Venus that the size of a photographic image varied according to the time of exposure, the intensity of the incoming light, and the aperture. See Wolf and Yvon-Villarceau, "Rapport sur les mesures micrométriques directes à faire pour l'observation du passage de Vénus, présenté à la Commission du passage de Vénus, dans la séance du 8 mars 1873," in *Recueil de mémoires, rapports et documents relatifs à l'observation du passage de Vénus sur le Soleil* (Paris: Firmin Didot, 1874), 339; Alfred Angot, "Bulletin des sociétés savantes, Académie des sciences de Paris séance du 22 Mai 1876: Les Photographies obtenues au foyer des lunettes astronomiques," *Revue scientifique* 17 (3 June 1876), and "Bulletin des sociétés savantes, Académie des sciences de Paris séance du 5 Juin 1876: Les Images photographiques obtenues au foyer des lunettes astronomiques," *Revue scientifique* 17 (17 June 1876). See also Jules Janssen, "Sur la constitution de la surface solaire et sur la photographie envisagée comme moyen de découverte en astronomie physique," *Comptes rendus des séances de l'Académie des sciences* 85 (31 December 1877): 1251–52.

15. Almost eight hundred photographs had to be measured. A machine to measure the plates was made by Brunner in November 1874 and finished in April 1875. This single machine, however, was not enough, and three others were built by the end of the year. One observer operated each machine. At first these were Cornu, Angot, Mercadier, and Baille. See Victor Puiseux "Remarques sur les observations du passage de Vénus du 8 décembre 1874," 27 March 1880, AAS, Carton 1646, Folder: 1882 Passage de Vénus, pp. 1–11; and Puiseux to Académie des sciences, Paris, 31 May 1880, AAS, Carton 1647, Folder: 1882 notes diverses.

graphic methods—and to some extent from astronomy itself—by advocating the determination of the solar parallax through measurements of the speed of light.

In 1881 a Conférence internationale du passage de Vénus was held, with the famed chemist and statesman Jean-Baptiste Dumas presiding.[16] The attendees asked, "Should we continue to employ photography, and to what degree?" Despite some scattered defenses of the technology, the overwhelming response was that "the photographic trials, taken as a whole, have cast a great incertitude on the value of the solar parallax." These discouraging results "led the French Commission to limit the use of photography." Scientists from around the world acknowledged that the transit of 1874 had greatly damaged the prestige of astronomers: "the scientific public was amazed to see that after seven years, there were only partial and few publications on the results of the observations of 1874."[17] By 1882 almost everybody agreed that nonphotographic observations were better.

Critiques of photography were most forcefully articulated by the German astronomer Wilhelm Förster, who called for the total elimination of photography for the 1882 transit. Ernest Mouchez, an admiral who had led one of the expeditions in 1874 and who would become director of the Paris Observatory, "agreed completely."[18] Förster found the probable error of photography versus direct micrometer measurements five times as large; he also described photographs that showed pentagonal images of Venus and others in which the planet appeared successively as "bitten" and then complete. Furthermore, the problem of photographic irradiation was added to that of scintillation; there were other "still unexplained" phenomena as well. Förster and his followers therefore relegated photography to its former pictorial function of providing images of the sun, moon, or star

16. Ministère de l'Instruction publique, "Première séance: Mercredi 5 octobre 1881," in *Conférence internationale du passage de Vénus* (Paris: Imprimerie Nationale, 1881), 4, Hervé Faye, "Association française pour l'avancement des sciences, congrès de Lille, conférences publiques: Le Prochain Passage de Vénus sur le Soleil," *Revue scientifique* 14, no. 16 (17 October 1874): 366; Ministère de l'Instruction publique, *Conférence internationale du passage de Vénus* (Paris: Imprimerie Nationale, 1881). The proceedings were reprinted in "La Conférence internationale du passage de Vénus," *Revue scientifique* 29 (14 January 1882).

17. Ministère de l'Instruction publique, "Première séance: Mercredi 5 octobre 1881," 10 (quoting Adolph Hirsch). For Hirsch's position, see letter from Hirsch, Paris, 12 October 1881, p. 1, and a note explaining the need for international cooperation and a single bureau of calculations for the transit of 1882, AAS, Carton 1645, Folder: Passage de Vénus papiers de J. B. Dumas, passage de Vénus 1882, pp. 1–3.

18. Dumas acknowledged that some photographs were very good. Emmanuel Liais was less hopeful because of photographic irradiation. Antoine d'Abbadie still threw his weight behind photography, acknowledging that the Americans under David Peck Todd had obtained good results. The British were generally against photography, since it was difficult to ask the government for more money after what had happened in 1874. They placed their hopes instead on the observation of "a considerable number" of contacts. Ibid., 7–8.

clusters. Besides the uncertainties posed by the photographs themselves, he complained about the "considerable" work needed to measure them—a problem that was only later solved in Paris by the incorporation of female labor in the observatory.[19] The overwhelming conclusion at the Conférence internationale was that direct observations were better than photographic ones.

Criticisms of photographic methods were widely shared. The astronomer Charles Wolf found the dream of "eliminating the observer" absurd. Observers, he claimed, would always be needed for obtaining "absolute and authentic knowledge." For Wolf, the eye's superiority resided in its stability across time. While different cameras and photographic processes produced different results (for example, collodion versus gelatin and bromide), "the human eye, on the contrary, is an organ which remains the same, and the observations of the eye are, at all times, comparable amongst themselves." Photography would never replace the (albeit expensive and sometimes dangerously revolutionary) observers employed in astronomy. After stating its benefits during a speech given to the Société française de photographie about the use of photography during the transit of Venus, he cautioned: "One should not believe, however, that photography will one day completely replace the observer."[20] In a paper written in 1886, he stressed how photography and direct observation would always complement each other: "The two types of observation complement each other. Both of them are necessary for attaining an absolute and authentic knowledge of the present state of the heavens."[21]

Another argument was based on the role of memory. Experienced observers could use their memory to their advantage, learning from accumulated impressions obtained through years of practice and distilling, in this manner, occurrences within short periods of time. An observer, Wolf explained, "stores in his memory the fugitive images he has been able to seize, and from the ensemble of impressions he summarizes the definitive image which he can reproduce in a drawing."[22] His views echoed with those

19. Charlotte Bigg, "Photography and the Labour History of Astrometry: The Carte du Ciel," in *The Role of Visual Representations in Astronomy*, ed. K. Hentschel and A. Wittman, Acta Historica Astronomiae (Thun: Deutsch, 2000).

20. Charles Wolf, "Conférence sur les applications de la photographie à l'astronomie et en particulier à l'observation du passage de Vénus," *Bulletin de la Société française de photographie* 21 (8 January 1875): 19.

21. Charles Wolf, "Sur la comparaison des résultats de l'observation astronomique directe avec ceux de l'inscription photographique," *Comptes rendus des séances de l'Académie des sciences* 102 (1 May 1886): 477.

22. Wolf, "Conférence sur les applications de la photographie à l'astronomie et en particulier à l'observation du passage de Vénus," 20.

of Förster, for whom the superiority of direct observation consisted in that, while instantaneous photography recorded only an instant, a good observer did a more valuable job by averaging over all instances.[23]

Wolf's superior, the astronomer Le Verrier, also combated photography. Although the official journal of the Académie des sciences often did not publicize divisions within it, alternative journals made a point of covering these debates.[24] Since 1859 he was allegedly overheard arguing that "the introduction of photography in astronomy is a bad idea, whose only effect will be to encumber science with deplorable errors."[25] The photographic methods the astronomer Hervé Faye planned to use for the solar eclipse of 18 July 1860, Le Verrier argued, "could not be used for truly scientific observations." The "très-vif et très piquant" debate between Le Verrier and Faye on photography led a commentator to the conclusion that even "when passing one's life contemplating the skies, one is not freed or detached from interests and terrestrial passions."[26] These confrontations, commencing well before the transit of Venus expeditions, only intensified afterward.

In the process of trying to determine what went wrong, scientists questioned their publication practices. Some argued to keep disagreements, difficulties, and errors hidden from the wider public and to only publish results once all controversies had been resolved. Why should they publicize their failed attempts to determine tenth-of-second moments? Certain attendees of the conference argued that "separate and hurried publications" on the upcoming transit should be prohibited and urged astronomers to "defer these until everyone had agreed." Through restraint, argued Förster, director of the Berlin Observatory, "the authority of astronomers would increase." Dumas also advocated a common publication to safeguard the "dignity of each country," and Förster was clearly being too frank when, thinking about the relation between the astronomers and the government, he said, "Scientific

---

23. Wolf, "Sur la comparaison des résultats de l'observation astronomique directe avec ceux de l'inscription photographique," 477. Wilhelm Förster explained: "a good observer overcomes this inferiority by fixing an average position from the images." D'Abbadie, however, turned this argument around, claiming that this visual averaging was a harmful characteristic, since because of it an observer might "neglect certain [fluctuating appearances] whose great importance might one day be demonstrated by the progress of science." Van de Sande Bakhuysen, director of the Leiden Observatory and representative of the Low Countries at the conference, remarked that the averaging advantage of the eye could be matched by photography by superimposing various images. Ministère de l'Instruction publique, "Première séance: Mercredi 5 octobre 1881," 6–7.

24. For a history of alternative scientific publications see Maurice Crosland, "Popular Science and the Arts: Challenges to Cultural Authority in France during the Second Empire," *British Journal for the History of Science* 34 (2001).

25. "Compte rendu des Académies. Bulletin de l'Académie des sciences. Séance du lundi 31 octobre 1859," *Journal du progrès des sciences médicales et de l'hydrothérapie rationelle* 4 (1859): 390.

26. Ibid.

liberty can be restrained a little, in order to assure a definitive result useful to the Governments who have a special right to it after having given extraordinary means." Not everyone agreed on the need for cooperation. Antoine d'Abbadie, one of the few defenders of photography present at the conference, opined that countries "should defend their liberty and publish their observations in their own manner," but the majority remained against him. Dumas argued that cooperation was "nothing extraordinary" but a "natural consequence of scientific evolution." "Before," he continued, "science progressed by the effort of isolated observers; later, the need for cooperation between savants of a same nation was felt, creating academies and national learned societies. Today, that is not enough, and one feels at all times the need for international gatherings of savants."

This internationalist attitude contrasted with the nationalist position taken earlier by most scientists, when different—incompatible—instruments were seen as an advantage, not as a problem. In 1874 the *Revue scientifique* had clearly noted that "because of the lack of resources and time, [France] cannot enter in a competition against the numerous expeditions of England, Russia and the U.S. We need, then, different procedures and different instruments to beat these rival nations."[27] Yet experience would eventually prove to most astronomers that international agreement and standardization with respect to instruments and methods was essential.

Standardization with regard to photography had not always been a concern for scientists. When Faye first presided over the Transit of Venus Commission, he and Janssen advocated the use of "a photographic instrument based on the same principles as that of the English, whose long experience had taught them the best methods." But after Faye left the group (allegedly because he "had other things to do"), "many members from the section of astronomy stopped going to the meetings."[28] Defenders of photography became increasingly torn, some arguing on behalf of sequential photography and the advantages of mechanically reproducible methods while others

27. Jules Janssen, "Discours prononcé à la première séance du congrès international de photographie, tenu à Paris, au palais Trocadéro, le 6 août 1889, par M. Janssen, président du comité d'organisation du congrès," in *Procès-verbaux et résolutions, du congrès international de photographie tenu à Paris du 6 au 17 août 1889* (1889); "Académie des sciences de Paris: Séance du 29 juin 1874," *Revue scientifique* 14, no. 1 (4 July 1874): 23.

28. Jules Janssen to the ministre de l'Instruction publique, 6 April 1876, p. 5, AN, F17 2928-2, Folder: Commission chargée d'examiner les comptes des dépenses de M. Janssen, nommée le 16 mars 1876, p. 5; and Jean-Baptiste Dumas and Elie de Beaumont (Secrétaires perpétuels de l'Académie) to the ministre de l'Instruction publique, Paris, 3 February 1873, AN, F17 2928-1, Folder C: Commission de l'Académie des sciences, travaux, préparations, etc., p. 2. After Delaunay's death, Faye became president of the commission on 1 September 1872. He explained the reasons for his departure and countered alternative theories in Faye to the Director of the revue, Passy, 3 April 1873, AAS, Carton 1645, Folder: Faye.

remained fixed on time-tested nonreproducible daguerreotypes. The Commission settled on daguerreotypes (silver-coated copper plates) instead of using collodion, the process chosen by most other nations and from which paper prints could be made. In the opinion of the Commission, the advantage of daguerreotypes' "conditions of inflexibility and invariability not offered by either paper or glass" outweighed the fact that they were not easily (mechanically) reproducible and that the images they captured might not be comparable to those of other nations. Faye and Janssen disagreed with the Commission's prescriptions on this issue.[29]

By 1882 almost every astronomer recognized, with Faye and Janssen, that in planning the 1874 transit expeditions they should have "agreed on the type of instruments and adopted everywhere the same dimensions in order to render observations more comparable."[30] The ministre de l'Instruction publique criticized how in 1874 "without prior agreement" nations had "acted in an independent and personal manner." Once again, Faye spoke out, expressing the hope "that the experience acquired at such a high price in 1874, should be useful in 1882, and that, this time, all civilized nations would unite their efforts in a common plan."

During a meeting at the Académie des sciences right before the ships left, Janssen showed his photographic results. The physicist Fizeau immediately protested the secrecy that had veiled them to that point: "Why had Janssen not communicated the result of his photographic research to the Commission of the Transit of Venus, of which he is a member and where he is involved with this issue?"[31] Although publicly "the *Compte-Rendu* said nothing" of criticisms against Janssen, widespread criticisms of photography included attacks on his methods. The astronomer was initially unable to convince his colleagues of the revolver's merits. In fact, the Commission did not adopt the device as its main instrument. In a private letter to the ministre de l'Instruction publique, Faye protested, pointing out the irony that

---

29. Adolphe Martin called attention to the superiority of daguerreotypes, and Yvon-Villarceau, Wolf, Cornu, and Fizeau all agreed. For a manuscript of Fizeau's work see Armand Fizeau, "Rapport sur l'appareil photographique," Paris, 8 March 1873, AAS, Carton 1645, Folder: Passage de Vénus, M. Fizeau. The instructions on the use of the photographic apparatus explain how during critical moments observers should take photographs every ten seconds. Förster disputed the superiority of daguerreotypes when compared to collodion. See Ministère de l'Instruction publique, "Première séance: Mercredi 5 octobre 1881," 6.

30. Ministère de l'Instruction publique, "Deuxième séance: Jeudi 6 octobre 1881," in *Conférence internationale du passage de Vénus* (Paris: Imprimerie nationale, 1881), 11, 17, 18 (Förster, Dumas, d'Abbadie), "Cinquième séance: Jeudi 13 Octobre 1881," in *Conférence internationale du passage de Vénus* (Paris: Imprimerie nationale, 1881), 26–7 (Dumas on cooperation), and "Première séance: Mercredi 5 octobre 1881," 4. One advantage of Janssen's revolver was that it had also been used in the expeditions of other nations.

31. "Académie des sciences: Séance du 6 juillet," *Le Moniteur scientifique* 16 (1874); "Académie des sciences.-6 juillet 1874," *Revue scientifique* 4 (1874).

an instrument invented by a Frenchman was adopted by other nations but was not authorized by the French Commission.

Janssen's public successes contrasted starkly with his standing in the eyes of the Commission. His colleagues did not forget that during his expedition to Japan he had taken matters into his own hands, ignoring official instructions. This resulted in the division of his mission into two separate expeditions, one to Kobe and the other to Nagasaki—and also to his boldly incurring extra expenses that exceeded his allocated funds.[32] Needing money to get home—and expecting to find allies in Paris—Janssen telegraphed Dumas asking for help. Dumas's response, however, made it clear that the academy would rather leave Janssen stranded in the East than extend further financial aid. In despair, Janssen asked, "How can we face our debts, finish our studies and return to France?" He was left with no option but to use his own funds.

Although Janssen disagreed with the conclusions of his colleagues, "a spirit of discipline" compelled him to employ the officially sanctioned methods during the expedition he was charged to lead to Japan.[33] Nevertheless, in addition to the officially prescribed instruments, he brought along delinquent ones, including his controversial "revolver."

The Problem of Synthesis

A particularly trenchant criticism haunting sequential photographic methods pertained to synthesis. Magic lantern technologies that created the illusion of movement by projecting drawings in rapid succession existed since the seventeenth century. By the 1830s various technologies used to produce the illusion of synthesis were improved by the Belgian physicist Joseph Plateau. Instruments that generated this sort of synthesis—such as thaumatropes, zoetropes, and phenakistiscopes—proliferated as the century progressed. The physicists Michael Faraday and James Clerk Maxwell made important improvements to these technologies. At the end of the century, techniques for displaying animated drawings had improved so much that Émile Reynaud, an expert in projection technologies, attracted thousands

32. Janssen's trip to Japan and his expenses were documented in his journal. See notebook for 1874, Bibliothèque de l'Institut de France, MS 4128. For a description of his plea for help see Janssen to ministre de l'Instruction publique, 6 April 1876, AN, F17 2928-2, Folder: Commission chargée d'examiner les comptes des dépenses de M. Janssen, nommée le 16 mars 1876, pp. 5–6. Regarding the reimbursement, see Chambre des députés, "Chambre des députés—Annexe no. 445: Séance du 3 août 1876," *Journal officiel de la République française* (14 September 1876).

33. Janssen to the ministre de l'Instruction publique, 6 April 1876, AN, F17 2928–2, Folder: Commission chargée d'examiner les comptes des dépenses de M. Janssen, nommée le 16 mars 1876, pp. 5–6.

of spectators on a weekly basis to his entertaining animations at the Musée
Grévin. Since his drawing-based spectacles were already immensely successful, Reynaud had only a marginal interest in photography. Scientists,
however, faced a different predicament. While in 1874 Janssen announced
that he had solved the "inverse problem of the phenakistiscope" with his
revolver, his solution introduced a new problem that would preoccupy scientists for the next quarter of the century: photographic synthesis.

In the decades preceding the invention of cinematography, scientists
were unable to use photographs to create moving pictures. Even when these
were coupled with magic lantern technologies the illusion of movement
was much less smooth than when drawings were used. Differences between
pictorial and photographic representations were stark when they were submitted to the test of magic lantern technologies. Even drawings made from
photographs worked much better. Critics claimed that the inability to use
photographs to obtain the effects of synthesis exposed an unstable epistemological fault line. Without synthesis, photographs remained unexplain-
ably different and even inferior to observational, artistic, and even popular
imaging techniques. The limitations of photography with respect to the
portrayal of movement remained particularly hurtful.

Methods of photographic synthesis were considered necessary "counter-
proofs" and "cross-checks" to the techniques of analysis created by Janssen.[34]
To obtain "an absolute proof of [photography's] truthfulness" scientists had
to improve on these technologies. Their scientific worth could only be established by "subjecting them to the test of the zoetrope."[35] Synthesis was the
"crosscheck used by all experimental sciences to verify the results of analysis, which is imposed in an absolute manner."[36] Combining magic lanterns,
phenakistiscopes, and zoetropes with photographs became a scientific im-
perative during the second half of the nineteenth century.

Janssen was unable to produce this synthesis. Marey, who since his first
incursions into photography started to search for synthesis, struggled with
this same difficulty. For more than half a century after its invention in 1839,
photography did not pass the test of synthesis. Scientists took and retook
photographs of varying times of exposure, points of view, and speeds of
succession, and they even experimented with cut and paste techniques. To
understand why they could not obtain synthesis they delved deep into the

34. P. Banet-Rivet, "La Représentation du mouvement et de la vie," *Revue des deux mondes* 40 (1 August
1907): 602, 05.
35. Edouard Cuyer, *Allures du cheval démontrées à l'aide d'une planche coloriée, découpée, superposée et
articulée* (Paris: J.-B. Baillière et fils, 1883), 14.
36. Banet-Rivet, "La Représentation du mouvement et de la vie," 602.

human body, fine-tuning their analyses of retinal impressions. They took into consideration elements such as nystagymus (involuntary ocular movements), retinal surfaces, individual variations, light intensity, background light, color, and having two, different eyes. Old, schematic analogies of the eye to the camera were revisited. Scientists looked beyond the retina. They experimented on reaction time and the speed of nerve transmission, investigated the role of mental and psychological factors in perceiving moving reality, and questioned common analogies between nerves and telegraphic wires. Photography's perceived inferiority in revealing movement, especially when compared to drawings, distanced it from the primeval source of empirical science: the senses.

A Return to Drawings

The inability to achieve the effects of synthesis with photography was part of a broader set of problems and critiques, some having to do with the relation between art and science, and others with the status of photography versus drawings. Drawing remained in the background of a vibrant culture of advancing photographic research. During this period direct observations and handmade drawings continued to be strong contenders for recording and visualizing short periods of time.

Harking back to older methods that were once discredited, a return to drawings was recommended at the 1881 International Conference on the transit of Venus. Observers were advised to "accompany their notes with a drawing."[37] Even photography's most hardened supporters agreed about the value of drawing methods. During this time, Janssen (one of photography's staunchest defenders) worked on two fronts, employing both photographers and draughtsmen.

During these years, photographic projects at major observatories increasingly turned to the production of great photographic atlases and catalogues. In 1886 Ernest Mouchez set in motion the gargantuan project of the *Carte du ciel* with the purpose of producing a map of the skies in collaboration with observatories all over the world.[38] Another important atlas focused on the moon. In 1894 Maurice Loewy (*sous-director* of the Paris Observatory)

37. Ministère de l'Instruction Publique, "Première séance: Mercredi 5 octobre 1881," 4, 6, and "Instructions pour l'observation des contacts," in *Conférence internationale du passage de Vénus* (Paris: Imprimerie nationale, 1881), 30 (drawings). The instructions for observing the contacts were reprinted in the *Revue scientifique* of 29 October 1881. Direct observations would be done in eight stations, and only in two would photography be employed.
38. Involving the collaboration of nations all over the world, the overly ambitious project dragged on into the twentieth century, only to be definitively abandoned in 1970.

and Pierre Puiseux (*astronome adjoint*) started working on a photographic atlas of the moon. The project lasted for fifteen years, published from 1896 to 1910. Janssen produced at Meudon a majestic photographic atlas of the sun, the first volume of which appeared in 1903.

Alternative photographic methods soon started to rival these atlas projects. In 1882, the year most astronomers turned against photography, Janssen urged Étienne Léopold Trouvelot, the most famous astronomical draughtsman at the time, to move from the Harvard Observatory to Meudon.[39] Trouvelot was particularly adept at depicting fugitive phenomena. By the time he moved to Meudon, his drawings were already famous—rivaling Lewis Rutherford's wet-plate photographs, as well as Warren de la Rue's and Janssen's own photographic methods.[40] In the years after he moved to Meudon, Trouvelot started questioning Janssen's photographic methods and volunteering alternatives based on moving cameras and lensless techniques.

When Janssen first met Trouvelot, he was impressed by "a series of drawings of planets which I thought were remarkable."[41] He "drew very well."[42] This judgment was widespread. Camille Flammarion, the famous astronomer and popular science writer, praised Trouvelot for "giving to

---

39. Trouvelot original manuscripts are kept at Houghton Library, Harvard University, and the Paris Observatory Archives, MS 1003-1006 bis. Documents on his affair with Janssen are at the Archives Nationales. All secondary accounts on Trouvelot overlook his photographic work. See E. Dorrit Hoffleit, "Trouvelot, Étienne Léopold," in *Dictionary of Scientific Biography*, ed. Charles Coulston Gillispie (New York: Scribner's, 1970); Jan K. Herman and Brenda G. Corbin, "Trouvelot: From Moths to Mars," *Sky and Telescope* 72 (1986); Brenda G. Corbin, "Étienne Léopold Trouvelot, 19th Century Artist and Astronomer," *Bulletin of the American Astronomical Society* 34 (2002); Françoise Launay, "Trouvelot à Meudon: Une 'affaire' et huit pastels," *L'Astronomie* 117 (2003); David Fossé, "Étienne-Léopold Trouvelot: Peintre du firmament," *Ciel et espace*, no. 421 (2005). For recent republications of his drawings, see Serge Brunier and Jean-Pierre Luminet, *Éclipses, les rendez-vous célestes* (Paris: Bordas, 1999), 24–25, 72, *Cosmos: Du Romantisme à l'avant-garde* (Paris: Gallimard, 1999), 51–54, *Trajectoires du rêve,* (Paris: Paris-Musées/Actes Sud, 2003), 126, 32–33, and cover. For exhibitions of his work, see *Cosmos: From Goya to de Chirico, from Friederich to Kiefer, Art in Search of Infinity* in the Palazzo Grassi in Venice; *Cosmos: Du Romantisme à l'avant-garde* at the Musée des beaux-arts de Montréal (1999); *Space 2001: To the Moon and Beyond* at the Bruce Museum of Greenwich, Connecticut; *Heavens Above* (2001) at the New York Public Library. Trouvelot's drawings were originally published in the *Annals of the Harvard College Observatory*, vols. 8 and 9; Étienne Léopold Trouvelot, *The Trouvelot Astronomical Drawings Manual* (New York: C. Scribner's Sons, 1882), and *Observations sur les planètes Vénus et Mercure* (Paris: Gauthier-Villars et fils, 1892).

40. For biographical details on Trouvelot written in the nineteenth century, see Étienne Léopold Trouvelot, "L. Trouvelot's Physical Observatory, Cambridge, Mass.," *Annual Record of Science and Industry for 1877* (1878), and "Cambridge, Mass.: Physical Observatory of L. Trouvelot, Esq.," *Annual Record of Science and Industry for 1878* (1879); "The Late M. Trouvelot," *Nature* 52 (1895); Émile Gautier, "Monsieur Trouvelot," *La Science française* 5, no. 14 (1895); "Nécrologie: E.-L. Trouvelot," *La Nature: Revue des sciences et de leurs applications aux arts et à l'industrie* (1895); "Étienne-Léopold Trouvelot," *Astrophysical Journal* 2 (1895).

41. Janssen, "Réponse du directeur Janssen," Archives Nationales de France (hereafter cited as AN), F17 3745, Folder: Observatoire de Meudon: Articles sur M. Trouvelot, pp. 1–8, on p. 1.

42. Janssen "Réponse du directeur Janssen," AN, F17 3745, Folder: Observatoire de Meudon: Articles sur M. Trouvelot, pp. 1–8, on p. 1.

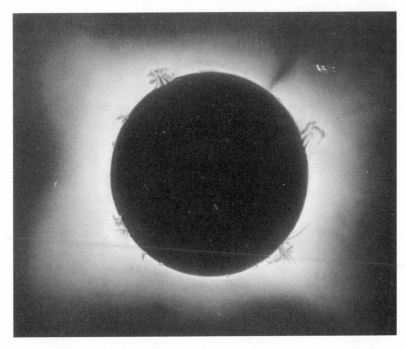

Figure 5.1. Pastel drawing of the sun with protuberances by Trouvelot (circa 1874). From the collections of the Paris Observatory.

Science the most beautiful series of telescopic drawings that exist."[43] An obituary at the time of his death claimed that his drawings were "known to all students of celestial science."[44] Janssen employed Trouvelot to observe and draw fugitive phenomena, initially charging him to "draw every day the solar protuberances with their forms, their dimensions, and their exact position on the sun's limb."[45]

43. Camille Flammarion, "Les Progrès de l'astronomie pendant l'année 1887," L'Astronomie 7 (1888): 163.
44. "The Late M. Trouvelot." Some of these drawings were exhibited at the International Exhibition of Philadelphia, and a selection of them was reproduced for pedagogical purposes. His career as a draughtsman was always tied to his career as a research scientist. He actively communicated his papers to the American Academy of Arts and Sciences, and published in the American Journal of Arts and Sciences and the prestigious Astronomische Nachrichten. His lithographs in the Annals of the Harvard College Observatory (1876, vol. 8, part 2, plates) represent some of its most important content. These lithographs were sometimes a product of personal observations, sometimes they were drawn from photographs, and other times were based on drawings by other observers. In all of these cases, they were considered important research work and sometimes tied to highly controversial debates. The topics of these papers included, but were not limited to, observations of the planets, clusters, nebulae, sunspots, comets, and solar protuberances.
45. Janssen "Réponse du directeur Janssen," AN, F17 3745, Folder: Observatoire de Meudon: Articles sur M. Trouvelot, pp. 1–8, on p. 1.

Figure 5.2. Solar prominences by Etienne Leopold Trouvelot. From *Annals of the Harvard College Observatory* (1876), vol. 8, part 2, plates.

Trouvelot's appetite for speed increased once he moved to Meudon. Yet, increasingly, he turned to methods for depicting fugitive events that were at odds with those used by Janssen and even those implemented in the Paris Observatory. Trouvelot's confrontation with state astronomers started as early as 1878, when Mouchez, as director of the Paris Observatory,

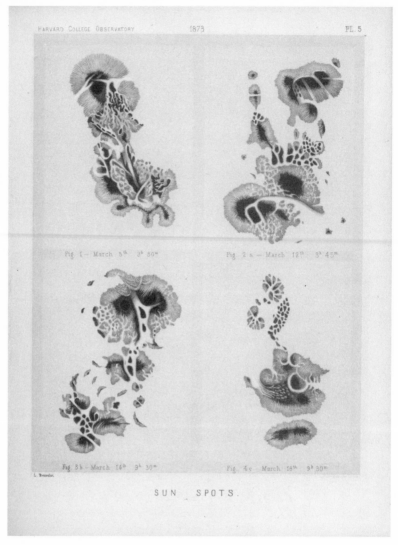

Figure 5.3. Sunspots by Etienne Leopold Trouvelot. From *Annals of the Harvard College Observatory* (1876), vol. 8, part 2, plates.

rejected a request to publish some of his images.[46] Their reproduction was "expensive," and Paul and Prosper Henry, two observers turned photographers who worked with him, were already producing comparable work. When an invitation finally did arrive from France, his situation did not improve.

46. Ernest Mouchez, "Astronomie," *Comptes rendus des séances de l'Académie des sciences* 87 (1878).

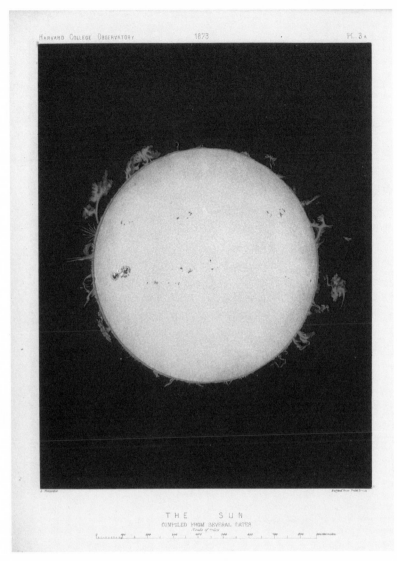

Figure 5.4. The sun showing protuberances and sunspots by Etienne Leopold Trouvelot. From *Annals of the Harvard College Observatory* (1876), vol. 8, part 2, plates.

In the years to come, Trouvelot's work fit less and less within the projects of the official observatories. He aspired to be a research scientist, and being an image maker for one of the major atlas projects under way in Meudon and Paris was hardly a road to fame and authorship. The photographs in these atlases were carefully presented as "authorless." When Janssen's heavyweight *Atlas de photographies solaires* was first published,

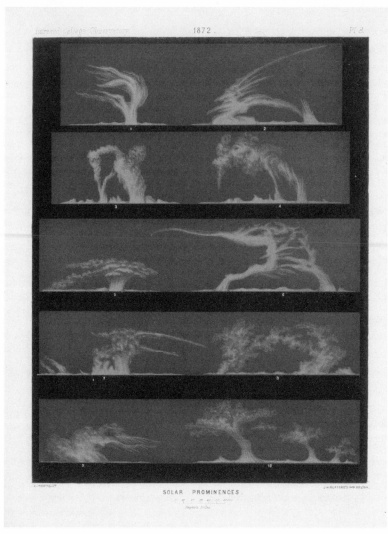

Figure 5.5. Sequence of solar protuberances by Etienne Leopold Trouvelot. From *Annals of the Harvard College Observatory* (1876), vol. 8, part 2, plates.

the author of the photographs was not even mentioned.[47] The assistant for the Paris Observatory's atlas of the moon, similarly, was not given authorship until the penultimate volume, which appeared after the death of Mau-

47. They were taken by Louis Pasteur (not to be confused with the famous doctor) with the assistance of Corroyer. Pasteur succeeded Arents. Louis Rabourdin also worked for Janssen. Françoise Launay, "Jules Janssen et la photographie," in *Dans le champ des étoiles: Les Photographes et le ciel 1850–2000* (Paris: Réunion des Musées Nationaux, 2000), 29.

rice Loewy, who had initiated the project.[48] Although Trouvelot disagreed with Janssen's attempts to portray astronomical photography as author-less, he started adopting the photographic technologies advocated by his superior. His use of photography, however, was completely different from those of the Meudon (and even the Paris) Observatory. Their differences soon became a major problem, causing a confrontation with his superior that eventually reached the public at large.

Trouvelot's choice of topics and methods lay in a productive zone with a long artistic and scientific tradition, that of depicting lightning and elec-tricity. Starting in 1888, he published numerous articles on these topics (appearing in *La Nature, L'Astronomie, La Lumière électrique,* and the pres-tigious *Comptes rendus* of the Académie des sciences). His images were then reproduced in "many other French and foreign scientific publications."[49] They touched on the disciplines of astronomy, physics, electricity, chemis-try, and even meteorology.[50]

In 1887 the Meteorological Society of London urged its members, many of them amateurs, to take as many photographs of lightning as possible.[51] That same year the British Association for the Advancement of Science advocated the creation of a "series of observatories for the regular and sys-tematic" photographic study of lightning.[52] In France, lightning and electric-ity photographs were pioneered by Charles Moussette.[53] Working from his

---

48. The assistant was Le Morvan.

49. Étienne Léopold Trouvelot, "La Photographie appliquée à l'étude de l'étincelle électrique," *La Nature: Revue des sciences et de leurs applications aux arts et à l'industrie* (1889): 110.

50. Étienne Léopold Trouvelot, "Étude sur la structure de l'éclair," *L'Astronomie* 7 (1888), "Étude sur la foudre," *La Lumiere électrique* 29 (1888), and "Étude sur la durée de l'éclair," *Comptes rendus des séances de l'Académie des sciences* 108 (1889).

51. For meteorological studies of lightning in Victorian science, see Jennifer Tucker, "Photography as Witness, Detective and Impostor: Visual Representation in Victorian Science," in *Victorian Sci-ence in Context,* ed. Bernard Lightman (Chicago: Chicago University Press, 1997), and *Nature Ex-posed: Photography as Eyewitness in Victorian Science* (Baltimore: Johns Hopkins University Press, 2005).

52. E. M., "Les Photographies d'éclairs," *La Lumiere électrique* 25, no. 37 (10 September 1887): 534.

53. An 1885 item by the American photographer William N. Jennings is—erroneously—usually consid-ered the first photograph of lightning. An earlier candidate is an 1847 daguerreotype taken by Thomas Easterly in St. Louis, Missouri, reproduced in Jon Darius, *Beyond Vision: One Hundred Historic Scientific Photographs* (Oxford: Oxford University Press, 1984), 25. Other early lightning photographers include the Austrian Robert Haensel (1883), reproduced in Gaston Tissandier, "Les Éclairs reproduits par la photogra-phie instantanée," *La Nature: Revue des sciences et de leurs applications aux arts et à l'industrie* (1884); J.-M. Eder, *La Photographie instantanée, son application aux arts et aux sciences,* trans. O. Campo, translated from the 2nd German edition. (Paris: Gauthier-Villars et fils, 1888), 120–21; Duquesne in Paris, 21 November 1884 at the Société française de physique and *La Nature* 1885, p. 32; Kayser in Berlin, Selinger, *Jahrbuch für photographie*; W. Prinz from Brussels; and Crow in England. By the turn of the century lightning photogra-phy was pursued on many fronts. See the review of this and other work on lightning in Em. Touchet, "La Forme et la structure de l'éclair," *La Nature: Revue des sciences et de leurs applications aux arts et à l'industrie* (1904). For a short history of early lightning photography, see Vladmir A. Rakov and Martin A. Uman, *Lightning: Physics and Effects* (Cambridge: Cambridge University Press, 2003), 3.

private observatory, he produced "very beautiful" photographs of lightning, electric sparks, and solar spectra.[54]

In the course of his research, Trouvelot noted how lightning sometimes seemed to last "many seconds." This effect was highly anomalous since most scientists agreed with the claim advanced by Charles Wheatstone, the inventor of the stereoscope, that lightning lasted less that a millionth of a second. Trouvelot boldly discounted the usual explanation that the much longer appearance of lightning was a mere visual illusion due to "the persistence of the luminous impression on the retina." He tried to disprove Wheatstone's theory by moving the camera during the time of exposure to produce a microscopic blurring effect.

From these experiments he concluded that the "instantaneity often attributed" to lightning was erroneous. He confronted Wheatstone directly: "The experiments of Wheatstone, which do not even attribute to lightning the speed of a thousandth of a second, are in error."[55] His conclusion was soon backed by the authority of Daniel Colladon, famous for having measured the speed of sound, who agreed that the velocity of lightning attributed by Wheatstone was too high.[56] Trouvelot surmised that the speed of lighting could only be calculated by constructing an "appropriate mechanism" that "communicated to the camera during the time of exposure a uniform, known movement." Although his publications did not contain any more information about the required mechanism, by using a camera in a way that was anathema to the atlas makers and chronophotographers, Trouvelot pioneered radically different methods for depicting movement. For this, he gained some allies. A correspondent to the *Comptes rendus* who cited his research applauded the experimental possibilities of "oscillating cameras."[57]

Research on electric sparks complemented his work on lightning.[58] Trouvelot joined forces with Eugène Ducretet, one of the most able construc-

54. Charles Moussette, "Orage du 12 mai 1886. La Foudre en spirale," *Comptes rendus des séances de l'Académie des sciences* (1886). His work was described and his images were reprinted in E. M. "Les Photographies d'éclairs."

55. Trouvelot, "Étude sur la durée de l'éclair," 1247. These experiments were followed by photographic images of a solar eclipse. By photographing the solar eclipse, he claimed to corroborate his previous observations of fugitive solar spectra. Étienne Léopold Trouvelot, "L'Éclipse de Soleil du 17 juin par M. E.-L. Trouvelot," *Comptes rendus des séances de l'Académie des sciences* (1890).

56. Daniel Colladon, *Comptes rendus des séances de l'Académie des sciences* 109 (1889). Colladon disputed Trouvelot's priority since he had made a similar point earlier—without, however, using photography.

57. Ch.-V. Zenger, *Comptes rendus des séances de l'Académie des sciences* 109 (1889): 295.

58. Photographs of electric sparks were taken by Rood; Pinaud; Schnauss; Federsen; Stein, *Das Licht im Dienste wissenschaftlicher Forschung* (Halle, 1886), 4:157; Van Melkebeke; Plücker; Welten; W. J. S. Lockyer; J. Brown; and the British engineer and entrepreneur William Armstrong (in 1897). Some of these photographs were reproduced in the French translation of the second German edition of Eder, *La Photographie*

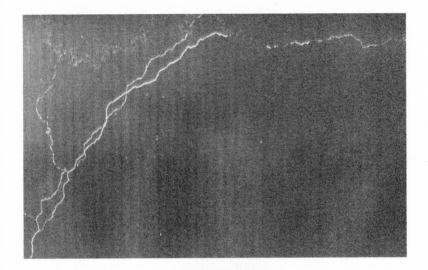

Figure 5.6. Reproduction of photographs of lightning by Trouvelot. The top image is a complete photograph from Étienne Léopold Trouvelot, "Étude sur la structure de l'éclair," *L'Astronomie* 7 (1888): 303–6, on p. 304, fig. 107. In the bottom image, Trouvelot amplified a six-millimeter portion of the complete image. By microscopically analyzing the image, Trouvelot placed limits on the speed of lighting. From Étienne Léopold Trouvelot, "Étude sur la foudre," *La Lumière électrique* 29 (1888): 254–55, on p. 254.

instantanée, son application aux arts et aux sciences. For Armstrong, see Chitra Ramalingam, "Visualizing Movement in William Armstrong's Images of Electric Discharge" (unpublished manuscript, Harvard University, January 2005).

tors of precision instruments and his associate Ernest Roger, both of whom would become famous for undertaking the first radio communications in Paris between the Eiffel Tower and the Pantheon.[59] In these experiments Trouvelot again employed photography in a radical manner—exploiting what most other photographers considered errors. His photographs of sparks were aperspectival and did not require a lens. They were produced by discharging an electric spark between two photographic plates stacked back to back.

Trouvelot's technique turned electricity into wild shapes. Some of his images resembled "large rivers, with their numerous tributaries, such as are reproduced in geographical maps"; others seemed like a "brushy hair," "small meteors," "firework bouquets," paradisiacal "palm trees" complete with thorns, biological "calyxes," and even "marine animals of the genre of the medusa."[60] The analogies used by him were both poetic and scientific. These picturesque, imaginative descriptions were aimed at disclosing real connections between diverse phenomena. Similarities were "so exact that one is led to believe that a botanist presented with the photograph of one of the extremities of certain branches would think it was a plant and not an electric phenomenon." His analogies included, most prosaically, plants, rivers, and fish, yet philosophically, they were used to question one of the sturdiest modern dichotomies with profound theological implications: the boundary between organized beings and unorganized bodies. Trouvelot posited a strong tie between "luminous phenomena and organized bodies."[61]

Some of these analogies were shared by Josef-Maria Eder, one of the first authors to publish photographs of lightning and electricity, who also marveled at their similarity. To him, thunderstorms revealed the spectacular work of a "celestial electrician." The similarities between lightning and electricity were a way for better understanding the work of an all-mighty

59. For a note on Eugène Ducretet's photographs of sparks, see Jamin, "Correspondance," Comptes rendus des séances de l'Académie des sciences 99 (1884).

60. Étienne Léopold Trouvelot, "Sur la forme des décharges électriques sur les plaques photographiques," La Lumière électrique 30 (1888): 270. The term "brushy hair" does not capture the meaning of the original "chevelure broussailleuse," where chevelure also refers to the tail of a comet.

61. Ibid. This article was based on an earlier communication, Étienne Léopold Trouvelot, "La Photographie appliquée à l'étude des décharges électriques," Comptes rendus des séances de l'Académie des sciences 107 (1888). These analogies became widespread. For J.-M. Eder: "The image of lightning resembles very much a geographic map in which the trace of a river receives the tributaries of other sources of running water." Eder, La Photographie instantanée, son application aux arts et aux sciences. A reviewer who compared Trouvelot's images against the photographs made by the British J. Brown in "conditions almost identical to M. Trouvelot's" described "palm leaves" and "leaves of numerous vegetables." H. W., "Figures produites sur des plaques photographiques par des décharges électriques, par J. Brown," La Lumière électrique 30, no. 51 (22 December 1888): 580. He referred to J. Brown, Philosophical Magazine 26, 502. For other accounts of Brown's work, see H. W., "Figures produites sur des plaques photographiques sèches par les décharges électriques, par J. Brown," La Lumière électrique 30, no. 42 (20 October 1888).

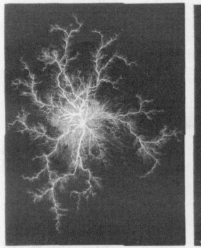 

Fig. 1. — Décharge positive d'une bobine de Ruhmkorff.   Fig. 2. — Décharge négative d'une bobine de Ruhmkorff.

Dans une communication faite l'année dernière à l'Académie [1], nous avons fait connaître une méthode à l'aide de laquelle on obtient, par le moyen de la photographie, des images directes de l'étincelle

Figure 5.7. Photographs of electric sparks. From Étienne Léopold Trouvelot, "La photographie appliquée à l'étude de l'étincelle électrique," *La Nature: Revue des sciences et de leurs applications aux arts et à l'industrie* (1889): 109–10, on p. 109.

source of power: "The celestial electrician uses the most powerful batteries and machines. It is not surprising that these sparks run for kilometers: that is their only difference."[62] To Eder's further amazement, this electrician had an eerie penchant for aiming at church steeples.

"Intelligent photography"

Trouvelot explained how his simple methods had "nonetheless inspired objections from many distinguished scientists."[63] One of those objecting was Janssen. One important disagreement between Janssen and Trouvelot hinged on the role of interpretation. Moving-camera images contrasted with static-camera photographs in that their interpretation was singularly difficult. The two men had further sources of disagreement. Janssen explained how Trouvelot gave a different name to his photographic method, calling it "intelligent photography." For Janssen this term implied that "ours was not intelligent, apparently." He mocked how Trouvelot "has searched in all sorts of directions: electricity, photography, etc. . . . and has not, in the end, left anything but tentatives which I have not been able to appreciate, fearing that I would not see in them their due value."[64] With time, the methodologies of Janssen and Trouvelot differed even further.

The collaboration between Janssen and Trouvelot quickly soured. The so-called Trouvelot Affair, closely followed by the independent media, centered on the premature death of Trouvelot. It was called a "scandal" by the press.[65] Debates about photography and drawings in astronomy were increasingly seen as political.[66] Politics, in fact, had always been central to Trouvelot. His move to America, and his residence there for more that three decades, had been the result of political exile because of his revolutionary activities against Napoleon III during the coup d'état of 1851. Once in Mas-

62. Eder, *La Photographie instantanée, son application aux arts et aux sciences*, 123.

63. Trouvelot, "Étude sur la durée de l'éclair," 1246.

64. Portions of the letter were reprinted in Launay, "Jules Janssen et la photographie." The citations are from pp. 459–60. Another objection came from Prinz, *Bulletin de l'Académie royale de Belgique* 16, 244. Reprinted in W. Prinz, "Étude de la structure des éclairs par la photographie," *La Lumière électrique* 30, no. 42 (20 October 1888).

65. Gaston Méry, "Une éclipse de Lune: Histoire de deux arbres qu'on n'a pas voulu abattre et d'un astronome qui en est mort," *La Libre Parole* (24 May 1895), 1–2; Gaston Méry, "A l'observatoire de Meudon," *La Libre Parole* (27 May 1895), 2 and "L'Observatoire de Meudon," *La Libre Parole* (29 May 1895), 2; G. M., "L'observatoire de Meudon," *La Libre Parole* (6 June 1895), 2; Émile Gautier, "Monsieur Trouvelot," *La Science française* 5, no. 14 (3 May 1895), and "Un Martyr de la science," *La Science française*, no. 19 (7 June 1895).

66. For a caution against facile readings of politics from technology, see Ken Alder, *Engineering the Revolution: Arms and Enlightenment in France, 1763–1815* (Princeton, N.J.: Princeton University Press, 1997), 293; Peter Galison, "Judgment against Objectivity," in *Picturing Science, Producing Art*, ed. Caroline A. Jones and Peter Galison (New York: Routledge, 1998), 355.

sachusetts, his new life was extraordinarily productive. He acquired the friendship of the famed naturalist Louis Agassiz and started illustrating natural history subjects and publishing in the *Annals* of the Boston Society of Natural History. Professor Joseph Winlock, director of the Harvard Observatory, impressed by his series of drawings, invited him to join. During his time at Harvard and from his private observatory in Cambridge, he produced the thousands of drawings which would make him world famous.[67]

Back in France, Trouvelot felt exploited. His family blamed the illness from which he died as brought on by overwork and by the repeated abuses by Janssen. Janssen explained the situation in a different way. Trouvelot was a man who felt that drawing fugitive events in the manner indicated by Janssen did not bring out "his personality." In response to the accusations leveled against him during the affair, the director of Meudon explained how Trouvelot neglected to do the job for which he was being paid: "It became clear that the regular observations which nevertheless constituted the basis of his work at the observatory and the reason for his remuneration displeased M. Trouvelot as not *bringing his personality into sufficient relief.*"[68] Janssen complained how his hired hand wanted money *and* authorship. Trouvelot, in turn, argued that Janssen recklessly spent the observatory's money precisely on projects that he considered to be inferior: photographic atlases.[69] Janssen believed that scientific images, especially those depicting fast events, had to be mechanical and impersonal. These divisions were resisted equally by Trouvelot and by other rogue members of the scientific community. Trouvelot nonetheless continued to work with photography despite the "frequently difficult circumstances" he faced with Janssen, but the two men would only remain close in the skies: remembered as the names given to craters on the moon and Mars.[70]

Gaston Méry, an outraged journalist working for the controversial journal *La Libre Parole,* disclosed the abuses taking place at the Paris and Meudon Observatories on various occasions.[71] Defenders who followed these affairs saw in them the injustices of the scientific establishment of the Third

---

67. For descriptions of his private observatory, see Étienne Léopold Trouvelot, "L. Trouvelot's Physical Observatory, Cambridge, Mass.," *Annual Record of Science and Industry for 1877* (1878), and "Cambridge, Mass.: Physical Observatory of L. Trouvelot, Esq.," *Annual Record of Science and Industry for 1878* (1879).

68. Janssen "Reponse du directeur Janssen," AN, F17 3745, Folder: Observatoire de Meudon: Articles sur M. Trouvelot, pp. 1–8, on p. 4.

69. Ibid.

70. "Nécrologie: E.-L. Trouvelot," *La Nature: Revue des sciences et de leurs applications aux arts et à l'industrie* (1895).

71. *La Libre Parole* was an anti-Semitic journal directed by Édouard Drumont. Gaston Méry was charged by Drumont to write in it a degrading article on Alfred Dreyfus during the famous Dreyfus Affair.

Republic. Janssen, a "member of the Institut [de France], member of the counsel for the order of the Legion of Honor, official scientist, and friend of [Gustave] Eiffel" stepped on the poor and defenseless.[72] It was a story of an "archidécoré"[73] scientist who "thought himself above the law and who followed no other rule of behavior other than his own pleasure."[74] One should be aware of these "Princes of Science" whose secret abuses were concealed by the "official press" and revealed only by brave "independent publications."[75] Disputes surrounding image making turned Meudon into "a nest of scandals."[76] Scientific photography was one of "scandals of all sorts—since there are others—for which the Meudon Observatory has been a theater for years."[77]

The journalist who exploited the "Trouvelot Affair" questioned astronomy's utilitarian connections. He underlined the complete uselessness of the Meudon Observatory. Although the Observatory was funded by the state and by taxpayers, Janssen refused to do a proper public accounting: "after seventeen or eighteen years of being founded" Meudon had not published "one single annual report, one *compte rendu*."[78] This taxpayer wanted results from state-sponsored observatories: "Well, we ask Janssen which are the discoveries, the studies, the observations done by him since the Observatory was installed in Meudon that justify this expense of three million?"[79] The expenses of the transit of Venus expeditions, the scientific "event of the century" where Janssen introduced his photographic methods, seemed to some similarly appalling. A mechanic working for the Transit of Venus Commission protested the government's funding of astronomy. He disparagingly compared the whole effort to "art for art's sake," reversing the orthodox conception that allied science with heteronomy and art with autonomy.[80]

The campaign against Janssen and defense of Trouvelot spread to other causes. In an "analogous story" to the one that had gripped Meudon, Méry defended Léon Jaubert, director of the Trocadero Popular Observatory. Symbolically inaugurated on Bastille Day, 14 July 1880, Jaubert's observatory

72. Janssen and Gustave Eiffel were close friends. Correspondance de Jules Janssen, Bibliothèque de l'Institut de France, MS 4135. Gaston Méry, "Une éclipse de Lune: Histoire de deux arbres qu'on n'a pas voulu abattre et d'un astronome qui en est mort," La Libre Parole (24 May 1895), 1–2.
73. *Journal officiel*, 26 May 1895, cited in G. M., "A l'observatoire de Meudon," La Libre Parole (27 May 1895), p. 2.
74. Gaston Méry, "A l'observatoire de Meudon," La Libre Parole (27 May 1895), 2.
75. Gaston Méry, "L'Observatoire de Meudon," La Libre Parole (6 June 1895), 2.
76. Gaston Méry, "A l'observatoire de Meudon," La Libre Parole (27 May 1895), 2.
77. Gaston Méry, "L'Observatoire de Meudon," La Libre Parole (6 June 1895), 2.
78. Ibid.
79. Gaston Méry, "L'Observatoire de Meudon," La Libre Parole (29 May 1895), 2.
80. Montagne and fils to ministre de l'Instruction publique, St. Etienne (Loire), 29 January 1874 and 9 April 1874, AAS, Carton 1645, Folder: Passages de Vénus, demandes de missions.

was an immense success, even threatening Meudon's supremacy.[81] Even Janssen's wife, Henrietta, thought it was prudent to warn her husband of the crowds flocking to the "so-called" observatory at Trocadero. "Tout Paris" wanted to be there.[82] Jaubert was yet another "persecuted scientist," first harassed by the director of the Paris Observatory, Urbain Le Verrier, and later hunted down by the next director, Mouchez. Powerful scientists in respected institutions, such as Le Verrier, Janssen, and Mouchez allegedly robbed labor, fame, and authorship from those at the bottom of the scientific food chain: "We have already recounted the history of M. Trouvelot, persecuted without respite by the friend of the developers of the Panama canal, the traveling astronomer Janssen."[83] Jaubert shared an essential characteristic with Trouvelot: he "drew very well."[84]

Scientists against Cinematography

The competition between the photographic methods used at major observatories and alternative techniques reached a climax in 1895, the year of Trouvelot's death and of the appearance of the Lumière cinematographic camera. Operating at 16–18 frames per second, the new camera showed how sequential images could be strung together to portray an illusion of movement. For the first time, the invention appeared to vindicate Janssen's methods of analysis. In its wake, the revolver reappeared as an important step in the development of the new medium. Trouvelot's speed photographs, in contrast, would be entirely forgotten. In the rare cases when he was remembered, he was portrayed as failing to keep up with photographic advances in science.[85]

---

81. Approximately three thousand people subscribed to it during its initial two years; almost four thousand visited it and used its instruments. Jaubert to the Ministre, 11 February 1882, AN, F17 2755, Folder: Observatoire populaire du Trocadero 1880–1882, no. 3, pp. 1–7, on pp. 3, 5.

82. Letter from Henrietta to Janssen, Oran, No. 222, Correspondance de Jules Janssen, Bibliothèque de l'Institut de France, MS 4134, 4 pp., on p. 1. The event in question was the second transit of Venus.

83. Gaston Méry, "Un Savant persécuté," *La Libre Parole* (10 July 1895). Copies of this article and one that followed it on 11 July 1895 appeared in *La Lanterne* and *L'Intransigeant*. A piece based on his second article appeared in *Le Journal*; see Launay, "Trouvelot à Meudon: Une 'affaire' et huit pastels," 456.

84. Loewy to the ministre de l'Instruction publique, 13 March 1891, AN, F17 2755, Folder: Jaubert, pp. 1–2, on p. 1.

85. To this day, Trouvelot's legacy is not marked by his astronomical work, but by his notorious experiments with silk production. While in Massachusetts, he raised millions of silkworms in the woodlands behind his house. In an unfortunate incident discreetly passed over in the loving obituaries detailed by his son, he decided to breed a native American insect with gypsy moths brought back from a trip to Europe. But the experiments went awry. Some moths escaped, and in the years that followed they caused the alarming defoliation of New England trees. In 1886 things took a turn for the worst. The entire neighborhood was infested; by the mid-1890s, the entire state. By then, however, he had moved back to France to work with Janssen. For the portrayal of Trouvelot as hopelessly anachronistic, see Jan K. Herman and Brenda G. Corbin, "Trouvelot: From Moths to Mars," *Sky and Telescope* 72 (1986).

Janssen presided over one of the first public demonstration of Lumière's apparatus, which took place in Lyon during a banquet for the congrès des Sociétés photographiques de France.[86] Not surprisingly some of the first films ever to be shown publicly were "movies" of Janssen himself. After showing the now famous *Promenade des congressistes sur les bords de la Saône,* where Janssen appeared prominently, the next movie was *M. Janssen, président du Congrès, discutant avec son ami Lagrange, conseiller général du Rhône* (12 June 1895).[87] Although these spectacles were still "tiring to see" since "the reproduced movements were jerky and accompanied by a regular fluttering which was extraordinarily disagreeable for the eye," an essential victory had taken place.[88] In the years after its invention, the cinematographic machine became a standard for recording, ending longstanding debates about the relation between analysis and synthesis.

Marey's fascination with the cinematographic camera did not reside in the fact that it showed moving images, but, rather, in seeing how it was used for both "synthesis" *and* "analysis." The "combination of photochronography with Plateau's apparatus" was essential for him.[89] Cinematographic machines were frequently described in terms of this connection: "All consist of a camera for obtaining the negative on a film, camera and mechanism which can serve to project the positive image."[90] Marey always employed only one machine.[91] He insisted on the benefits of having *one* apparatus perform *two* functions: "The chronophotographe does not need any modifications to become a projector."[92] Differing from Muybridge, his American

86. See journal entry for Tuesday 11 July 1895 describing Janssen's arrival at Lyon at 4:26 in the morning, and a rendezvous at the embarcadero at 7:30. The entry for 12 July 1895 says "je montre ma plaque Daguerriene du revolver." After having dinner with the Lumière brothers, Janssen lists "projections phot." Correspondance de Jules Janssen, Bibliothèque de l'Institut de France, MS 4131, carnet 7.

87. The well-known first comic film *L'Arroseur arrosé* followed Janssen's. Marta Braun, *Picturing Time: The Work of Etienne-Jules Marey (1830–1904)* (Chicago: University of Chicago Press 1992), 194–95. Right after Lumière's "movies" were shown in Lyon, Janssen was chosen as the speaker at the congrès de l'Union nationale des Sociétés photographiques. Janssen's speech to the Union nationale des Sociétés photographiques de France, 12 June 1895, was reported in the *Bulletin de la Société française de photographie* 11 (1895): 423. Michel G. Coissac, *Histoire du cinématographe* (Paris: Éditions du "Cinéopse," 1925), 247.

88. Georges Vitoux, *La Photographie du mouvement* (Paris: Chamuel, 1896), 16–17.

89. See Marey, "La Photochronographie et ses applications à l'analyse des phénomènes physiologiques," 516.

90. This connection is explained in Eugène Trutat, *La Photographie animée,* Bibliothèque photographique (Paris: Gauthier-Villars, 1899); Ch. Fabre, *Les Industries photographiques,* Encyclopédie industrielle (Paris: Gauthier-Villars), 512. An important exception is Georges Demenÿ, who used two machines, one for taking the photographs, and the other one for showing them. He used a disk instead of film, so was only able to reproduce short or periodic movements.

91. A book on the "photographic industry" remarked on the essential trademarks of Marey's machines: "In Marey's apparatus, as in the majority of similar instruments, the same instrument serves to produce the negative images and to project the positive one." Fabre, *Les Industries photographiques,* 512.

92. Marey, "Nouvelles modifications du chronophotographe par M. E. J. Marey," in Louis Gastine, *La Chronophotographie, encyclopédie scientifique des aide-mémoire* (Paris: Gauthier-Villars et fils), 69.

Figure 5.8. Janssen in a Lumière film. From Marta Braun, *Picturing Time: The Work of Etienne-Jules Marey (1830–1904)* (Chicago: University of Chicago Press, 1992).

predecessor, who used many cameras, he insisted that all images be taken with a single lens, demanding constant intervals between them when both recording and projecting. Additionally, photographs had to be left in the same sequence in which they were taken and remain unedited.[93]

93. "Photochronography" was Marey's preferred term because it underlined the equal time intervals that differentiated his method from Muybridge's. At the congrès photographique de 1889 the term "chrono-photography" was adopted.

What most scientists like Marey sought in their cinematographic investigations was a *connection* between photographic analysis and synthesis. The governing impetus behind devising cinematographic technologies—that is, for achieving with photography the synthesis that had long been produced with drawings—was driven by this need, as was the growing interest in cinematography as a form of recording and the related modern condition that would emerge as cinephilia.

The connection between shooting and projecting largely lost its raison d'être by the end of the century. In 1897 the Lumière brothers produced a cheaper camera designed only for projecting. Ten years later a writer to the popular *Revue scientifique* remarked on the necessity of separating the recording and projecting functions of the apparatus: "industrial practices have shown that instead of *inverting the chronophotograph,* it is preferable to separate the instruments from each other. That is to say, to construct special machines for photographic recording."[94] These changes had important consequences. Having two machines, instead of one, widened the gap between spectators and producers. Another change involved increasingly hiding the machine's internal mechanisms. While most scientists strove to make the mechanisms regulating synthesis and analysis as visible and evident as possible, the cinematographic industry was concerned with exactly the opposite in order to enhance the medium's phantasmagoric effect.[95]

When "industrial practices" revealed the advantages of separating the instruments for recording from those for projecting, some scientists felt that their efforts to establish a cinematographic connection between synthesis and analysis had been subverted. They saw, just a few years after their decades-long search came to an end, their solution fall apart. Many scientists, including Marey, resisted these changes. They explained to the public what they considered to be the essence of cinematography and stressed *their* role in its invention. While many scientists, including numerous astronomers, dedicated a great part of their lives to "inverting the chronophotograph," once this inversion was completed the public found no need for it. Scientists, including Janssen and Marey, who had made so much progress with chronophotography, suddenly left the development of cinema-

94. Banet-Rivet, "La Représentation du mouvement et de la vie," 614. Italics mine.
95. For the connection between illusion, phantasmagoria, and "the occultation of production by means of the outward appearance of the product," see Theodor Adorno, *In Search of Wagner,* trans. Rodney Livingstone (London: New Left Books, 1981). And its discussion in Jonathan Crary, *Suspensions of Perception: Attention, Spectacle, and Modern Culture* (Cambridge, Mass.: MIT Press, 1999), 261. For the concern of scientists with making the magic lantern's mechanism visible, see Anne Secord, "Botany on a Plate: Pleasure and the Power of Pictures in Promoting Early Nineteenth-Century Scientific Knowledge," *Isis* 93 (2002).

tography to others and frequently opposed further improvements to the medium.[96]

Editing became common. Movies fascinated and entertained. Other changes followed, such as introducing sound. Cinematographic technologies moved out of the laboratory and into places of spectacle. Marey lamented changes in this direction. By 1899 he had become painfully aware that the reason why photography "has gained popularity, is not due to its true worth [*valeur véritable*]: she has gained this good fortune by interesting the public with the charming illusions which she gives."[97] The physiologist did not believe the public's fascination could be sustained: "No matter how perfect is the reproduction of familiar scenes, we start growing tired of watching them. The animation of a street with passersby, the horses, the vehicles which cross each other in diverse directions do not suffice anymore to capture attention." Even if new tricks were found, he believed they would eventually tire the public: "Already the search of curious subjects is imposed. One demands of far away places new spectacles, which, soon, will not be sufficient in themselves for sustaining interest." He was wrong.

Marey distanced himself from these spectacular trends. In numerous publications, he described cinema's faults and chronophotography's merits by differentiating chronophotography from the vulgar vision-based spectacles of cinema: "Chronophotography must, then, renounce the representation of phenomena as we see it."[98] In his *Traité de chronographie* (1899) describing Lumière's methods, he was characteristically deprecatory: "Furthermore, the absolute perfection of projections which naturally provoke the public's enthusiasm, is not that which, personally, concerns me the most." One year later, for the 1900 Universal Exhibition in Paris, he continued to insist: "Animated projections, interesting as they are, are of little advantage to science, for they only show what we see better with our

96. The historian of photography Joel Snyder describes Marey's paradoxical position as a "wobble" in Joel Snyder, "Visualization and Visibility," in *Picturing Science, Producing Art*, ed. Caroline A. Jones and Peter Galison (New York: Routledge, 1998), 393.

97. Marey, Préface in Trutat, *La Photographie animée*, v.

98. The complete statement is "but in the end that which [animated photography] shows, the eye can see directly. They add nothing to the power of our eyes, they do not remove any of their illusions. And yet, the true character of the scientific method is to supplement the insufficiency of our senses or to correct their errors. To achieve that, *Chronophotographie* must, then, renounce the representation of phenomena as we see it." Ibid., vii. He expressed this same idea again to his fellow physiologists: "But the true utility of the phenakistiscope is not to give us the same sensation which our eyes perceive in the face of moving objects, it should show us that which our eyes cannot perceive through direct observation, that is to say, it should detail all the phases of an act which seems confused due to the persistence of retinal impressions." Marey, "La Photochronographie et ses applications à l'analyse des phénomènes physiologiques," 513–14.

own eyes."[99] They were only interesting, according to him, when they were significantly slowed down or accelerated rapidly.

Leaving the Laboratory

The essence of cinema was increasingly defined in terms of its emancipation from the laboratory—a step that occurred simultaneously with the adoption of filmstrips.[100] The development of the cinematographic industry was part of a modern trend that extended laboratory methods to the outside world through new industrialized and standardized production methods. Eugène Trutat, honorary president of the Société photographique and director of the Muséum d'histoire naturelle at Toulouse noticed how "from the laboratory of men of science the modern discovery goes to the factory."[101] This process was often welcomed and driven by scientists, who saw in it a way of expanding laboratory knowledge into the outside world through standardized large-scale networks.[102] In the case of cinematography, however, the move to the factory produced a different effect. Instead of extending its scientific functions outward, industrialization distanced it from science, especially once the recording and projecting functions were separated.

Even when it was undeniable that the "valeur véritable" of cinema as conceived by him was not fully appreciated, Marey still believed the allure of spectacle would one day end. Chronophotography and its "inverse" would return to the privacy of the laboratory: "It is then that Chronophotography, returning to its origins, will become scientific once again. She will become the vulgarizer of these always novel spectacles, always captivating, which savants still enjoy alone in their laboratories."[103] Although a number

99. Étienne-Jules Marey, "The History of Chronophotography," *Annual Report of the Board of Regents of the Smithsonian Institution, Showing the Operations, Expenditures, and Condition of the Institution for the Year Ending June 30, 1901* (1902): 329. Translation of Marey, "Exposition d'instruments et d'images relatifs à l'histoire de la chronophotographie," in Musée Centennal de la classe 12 (Photographie) à l'Exposition universelle internationale de 1900 à Paris—Métrophotographie et chronophotographie.

100. The filmstrip was the essential characteristic of cinema according to Trutat, *La Photographie animée*, 35, 37–38, 42.

101. Eugène Trutat, *Dix leçons de photographie. Cours professé au Muséum de Toulouse*, Bibliothèque photographique (Paris: Gauthier-Villars, 1899). For the relationship of the artistic studio, the laboratory, and the factory, see Hans Ulrich Obrist and Barbara Vanderlinden, eds., *Laboratorium* (Cologne: Dumont, 2001); and especially, Peter Galison and Caroline A. Jones, "Trajectories of Production: Laboratories/Factories/Studios," in *Laboratorium*, ed. Hans Ulrich Obrist and Barbara Vanderlinden (Cologne: Dumont, 2001).

102. Simon Schaffer, "Late Victorian Metrology and Its Instrumentation: A Manufactory of Ohms," in *Invisible Connections, Instruments, Institutions and Science*, ed. Robert Bud and Susan Cozzens (Bellingham, Wash.: SPIE Optical Engineering Press, 1992).

103. Marey, Préface in Trutat, *La Photographie animée*, vi.

of scientists continued to use film in laboratory settings, cinematography never returned completely to the laboratory.

New criticisms started to be launched against cinema. While scientists celebrated the solution to a decades-long problem, their elation was short-lived. Astronomers during this period encountered new problems with cinematography. The different shooting and projecting instruments and speeds of commercial cinematography made it largely useless for obtaining precision measurements.

By the first decades of the twentieth century astronomers placed their hopes on noncinematographic instruments for recording transits. In 1910 an "impersonal micrometer" was adopted in the Paris Observatory for observing stars moving across the sky.[104] With a "traveling wire" that bisected the image of a star, it recorded its movement automatically. Even Faye, who had been earlier enthusiastic about photography's ability to replace the observer, eventually changed his expectations, placing his trust on the new micrometer: "The only resort, in order not to be arrested at the actual level of precision, is to eliminate the observer as has been more or less successfully accomplished using the impersonal micrometer."[105]

For recording the solar eclipse of 17 April 1912 and the lunar eclipse on 12 March 1914 astronomers were forced to film a clock next to the celestial phenomena in order to adequately keep track of the passage of time.[106] Because of these difficulties, most astronomers simply abandoned the

104. The most successful "impersonal micrometer" was built by Repsold. It was introduced in 1889 by the Repsolds in Munich to determine star transits. Repsold's micrometer had a hand-driven mechanism that would follow the star's constant motion. Dr. Gill in 1896 tried a hand-driven Repsold micrometer at their shop in Munich but found the effort to follow the star fatiguing and expressed that it could be improved by driving the wire mechanically. An "impersonal micrometer" driven by an electric motor was put in use at the Cape Observatory in 1905, another one was installed in the Paris Observatory in 1910, and in Greenwich a hand-driven micrometer was adapted to the old transit circle in 1915. Sir Charles Wheatstone first recommended the use of a "traveling wire" in 1864. His idea was developed by Rédier of the Paris Observatory. Before Wheatstone, Dr. Carl Braun introduced a similar machine in 1861. Repsold's method was described in *Astronomische Nachrichten*, no. 2940 (1889) and in François Gonnessiat, *Recherches sur l'équation personnelle dans les observations astronomiques de passage*, Annales de l'université de Lyon (Paris: G. Masson, 1892), 3:163–64. Also see Radau, "Sur les erreurs personnelles," 1028–30; Henri Renan, "Le nouveau micromètre enregistreur du cercle méridien du jardin de l'Observatoire de Paris," *Annales de l'Observatoire impérial de Paris* 26 (1910); Walter W. Bryant, *A History of Astronomy* (London: Methuen & Co., 1907); Duncombe, "Personal Equation in Astronomy"; Anton Pannekoek, *A History of Astronomy* (London: G. Allen & Unwin, 1961).

105. Hervé Faye, *Cours d'astronomie et de géodésie de l'École polytechnique* (Paris: Gauthier-Villars, 1883), 160. In a previous edition Faye's remarks did not yet include the "impersonal micrometer." He curtly wrote: "The only resort, in order not to be arrested at the actual level of precision, is to eliminate the observer." Hervé Faye, *Cours d'astronomie de l'École polytechnique*, 2 vols. (Paris: Gauthier-Villars, 1881), 2:164. Despite these expectations, a bisection error, similar to the one found in photographic and graphic methods, remained. Additionally, a new type of error, called leading error, where different observers tended to drive the micrometer a little ahead or behind the star's position, affected the new instrument.

106. Fred Vlès, *La Cinématographie astronomique*, Bibliothèque générale de cinématographie (Paris: Charles Mendel, c. 1914).

medium. In physiology, the medium underwent a similar transformation. In 1911, the Institut Marey considered its future as a "center of cinematography applied to physiology," but it was hampered by "chronophotography's only inconvenience of being very expensive." The film production company Pathé offered to furnish the Institut on a yearly basis with "a considerable number of meters of film," on the condition to have in exchange the monopoly for exploiting all of their films. After "an important discussion" this proposition "was not accepted, because of the danger that the Institut could suffer by the public projection of certain sensational experiments."[107] The rejection of Pathé's generous offer signaled an important change from early cinematographic work that was simultaneously at the cutting edge of both science and spectacle.

By midcentury the distance between scientists and cinematographers had widened to the point that the film critic André Bazin chastised the preeminent historian of cinema Georges Sadoul, who briefly mentioned Janssen in his famous *Histoire du cinéma mondial des origines à nos jours*, for tracing the origins of cinema to science.[108] In contrast to Sadoul, Bazin claimed that "the cinema owes virtually nothing to the scientific spirit."[109]

The inability to perceive moments of the order of a tenth of a second led scientists to search for technologies to overcome these limitations. Photography first appeared as a favorite candidate, and numerous scientists explicitly cited the tenth of a second in their defense of the technology. Yet photography, even of the high-speed kind, was soon perceived to be deficient by many astronomers. Although it captured short slices of time, frequently going beyond tenth-of-a-second limits, it seemed (in the eyes of its critics) to lose its ability to capture movement. For this reason, photography continued to compete against drawings for more than half of a century after it was first invented. Diverse imaging techniques, some drawing-based, others photographic, some lensless, and others based on moving cameras, all claimed to capture movement. For years, no clear consensus emerged as to the benefits of one single technique.

Faced with a diversity of methods and rampant disagreement on the merits of each, many scientists realized that tenth-of-a-second limitations showed the need to explore not only photographic but cinematographic technologies. In 1895, the year commonly considered to mark the invention of cinematography, they celebrated the new technology. Yet their ela-

---

107. Notebook, Procès verbaux des réunions, Marey Box CXII, Collège de France.
108. Georges Sadoul, *Histoire du cinéma mondial des origines à nos jours* (Paris: Denöel, 1946).
109. André Bazin, "The Myth of Total Cinema," in *What Is Cinema?* ed. Hugh Gray (Berkeley: University of California Press, 2005 [1946]), 17.

tion was short-lived. Cinematographic technologies soon escaped scientific laboratories to become, primarily, technologies of entertainment. While some of the first cinematographic cameras used by scientists "recorded" and "played back" events at the same speed and using a single machine, later cameras separated these functions, making it increasingly hard for scientists to use them as a way of studying temporal events with precision. Scientists searched for a cinematographic solution and defended the adoption of the medium by reference to the tenth of a second, yet, despite their decades-long efforts, they saw its importance diminish in science and expand in other areas of culture.

But even with these changes and despite the disappointment that surrounded the new technology, the tenth of a second nonetheless entered the cinematographic era—an era marked by important changes in how scientists, philosophers, and cultural critics thought about these moments and their relation to movement.

CHAPTER 6

STABILIZING PHYSICS

The determination of time . . . happens at the order of 1/10 of a second; the
persistence of retinal impressions is of the same order.

ALFRED CORNU, physicist

By the turn of the century the tenth of a second appeared at the center
of some of the most important debates in physics. In heated discussions
about the advantages and disadvantages of the prevailing division of time
in terms of day, hours, minutes, and seconds, this unit appeared tied to
nature as no other.[1] The physicist Alfred Cornu, professor of physics at
the École polytechnique, explained how our capacity to measure time was
limited by the same tenth-of-a-second intervals that marked visual persistence effects: "The determination of time . . . happens at the order of 1/10
of a second; the persistence of retinal impressions is of the same order."
This unit was a "sort of physiological constant" and a "privileged magnitude
in harmony with the human organism" essential for obtaining precision:
"We should not stray far from it," warned the physicist, unless scientists

---

1. For the context of these debates see Peter Galison, *Einstein's Clocks, Poincaré's Maps: Empires of Time*
(New York: W. W. Norton and Co., 2003), 170–71.

wanted to "lose the precision which gives time measurement such a considerable value."[2]

Cornu arrived at these conclusions while measuring the speed of light. These experiments led him to study the phenomenon of visual persistence, where "the eye does not distinguish individually a series of identical perceptions which are reproduced periodically if the period of their succession becomes less than approximately a tenth of a second."[3] He also studied another delay that would "rarely reach a 1/10 of a second" and that appeared in "almost the same form" as in the "determinations of star transits across meridian instruments" when telegraphic instruments were used for establishing time and longitude.[4] His results depended on these essential physiological properties and on their relation to the tenth of a second.

The problems brought about by the tenth of a second motivated some of the most important changes in modern physics. Speed-of-light experiments in physics gained importance because of tenth-of-a-second errors in astronomy. Interferometry, a technique considered to be at the pinnacle of precision science, was developed to evade errors of this magnitude. A new definition of the meter (in terms of light waves) gained currency because of tenth-of-a-second errors that haunted previous determinations of this standard.

Cornu started his career after graduating second in his class at the École polytechnique. In 1871 he began experimenting on the speed of light. His experiment consisted in bouncing a source of light off a mirror and intercepting it at its return with a spinning toothed wheel. By varying the speed of the rotating wheel, and noting the appearance and disappearance of the source of light through the wheel's teeth, he was able to calculate the speed of light. He explained how, in measuring "the total duration of such a bright luminous impression on the retina," "one takes it to be less than 1/10 of a second."[5] A viewer in front of the rotating wheel would see fusion effects at the ocular when "ten teeth passed in a second." At this exact velocity the "eye of the observer would not distinguish the absence or the existence of the teeth, nor interruptions in the source of light: he will see a fixed point of light, caused by the persistence of impressions on the retina."[6]

2. Alfred Cornu, "La Décimalisation de l'heure et de la circonférence," *L'Éclairage électrique* 11 (1897): 388.

3. Cornu referred to the research on this topic presented by Helmholtz in his *Physiological Optics*. He cited Lissajous's experiments as proof that it could be lowered to one-thirtieth. Alfred Cornu, *Détermination de la vitesse de la lumière d'après des expériences exécutées en 1874 entre l'observatoire et Montlhéry, extrait des Annales de l'Observatoire de Paris (Mémoires, tome XIII)* (Paris: Gauthier-Villars, 1876), 7 n.

4. Ibid., 101.

5. Ibid.

6. Ibid.

The physicist instructed the observer to mark the moment when this effect surfaced.

Cornu's investigations had a number of repercussions. Astronomers used speed-of-light measurements to calculate a value for the solar parallax so dearly sought during the transits of Venus. In his famous publication, he explained: "These experiments have a truly current importance since they permit us to determine with exactitude the value of the solar parallax, which astronomers of all nations are demanding from the next transit of Venus across the sun at the price of costly voyages, both difficult and risky."[7] For metrologists, this value promised to be a sturdier standard of measurement than preexisting ones.[8] Speed-of-light determinations threatened to alter longstanding hierarchies between physics and astronomy, in that they showed how physicists, and not only astronomers, could contribute to the establishment of important constants of celestial mechanics. Speed-of-light measurements would take precision science to the next frontier.

This chapter reevaluates the history of speed-of-light measurements by tracing their trajectory from the work of Cornu to that of the famous physicist Albert A. Michelson. While these experiments have generally been considered only in light of later developments in modern physics, this chapter argues for a different historical and philosophical genealogy. Examined through the lens of the tenth of a second, the rise of physics' reputation as the exemplar of precision science looks different. A new era in science saw a rise in the status of physics vis-à-vis astronomy and the proliferation of metrological institutes. It was marked by a decline in Humboldtian-style expeditions designed to find precision constants, by the rise in networked

---

7. Alfred Cornu, "Détermination nouvelle de la vitesse de la lumière," *Journal de l'École polytechnique* 27 (1874): 139. Part of this work was reprinted in Alfred Cornu, "Détermination nouvelle de la vitesse de la lumière," *Comptes rendus des séances de l'Académie des Sciences* 76 (1873). After the presentation of Cornu's work to the *Académie des sciences,* Le Verrier insisted on sending it to the transit of Venus commission. Le Verrier, Paris, 21 May 1875, AN, F17 2928-1, Folder C: Commission de l'Académie des sciences, travaux, préparations, etc. (documents supporting Cornu's experiments on the speed of light); Note to ministre de l'Instruction publique, 15 May 1875, AN, F17 2928-1, Folder C: Commission de l'Académie des sciences, travaux, préparations, etc., (asking whether Cornu's new petition for funds to determine the speed of light should fall within the budget for the transit of Venus); and Cornu, Courtenay (Loiret), 15 March 1875, AN, F17 2928-1, Folder C: Commission de l'Académie des sciences, travaux, préparations, etc., pp. 1–2 (asking for money for his experiments and saying that "since the *main interest in the direct determination of the speed of light is due to the computation of the solar parallax,* we were forced to finish the instruments, do the experiments, and publish the definite result before the middle of December 1874"; he also described how his value "conformed to the results of M. Le Verrier." Italics mine). Cornu to ministre de l'Instruction publique, 5 February 1875, AN, F17 2928-1, Folder C: Commission de l'Académie des sciences, travaux, préparations, etc. (Cornu, supported by Armand Fizeau and Urbain Le Verrier, asked for money to pay Louis Bréguet, member of the renowned family of clockmakers, for his services).

8. For an account of speed-of-light measurements and their relation to the problem of standards see Richard Staley, "Traveling Light," in *Instruments, Travel and Science: Itineraries of Precision from the Seventeenth to the Twentieth Century,* ed. Marie-Noëlle Bourget, Christian Licoppe, and H. Otto Sibum (London: Routledge, 2002).

colonial relations, and by a change in the relationship between astronomy and physics.

Light Signals

Cornu's research on personal errors in speed-of-light measurements led him to reevaluate the work of his predecessors in these areas. The two principal scientists known for their speed-of-light measurements were Armand Fizeau and Léon Foucault, who were once close friends, but who eventually fought over various issues, most importantly over the wave or particle nature of light.[9] Going back to the seventeenth century, this debate involved figures of the stature of Newton, who argued for its particle nature. Projecting forward to the twentieth century, it would become central to the foundations of quantum theory, when its dual wave-particle nature was established. Fizeau and Foucault's experiments of the mid-nineteenth century proved to most scientists that light traveled faster through air than through water. This effect was widely interpreted as confirming the wave theory of light. While working together in the experiment, Fizeau and Foucault became engaged in a bitter dispute from which their once intimate friendship would never recover. But despite the disagreement between the two experimenters, the verdict for these highly public experiments fell in favor of the wave theory of light.

Fizeau and Foucault's experiments on the speed of light were first instigated by François Arago, who, suffering from old age and faltering eyesight, charged Fizeau, one of his best students, with the task of measuring the near-infinite velocities of light and electricity. Fizeau had already made a name for himself in science by working with Arago on various other projects, including astronomical photography. He paired up with Foucault, who started his career as a microscopy assistant interested in daguerreotypes and who would obtain fame through his famous "Foucault's pendulum" experiment elegantly demonstrating the speed of the earth's rotation. In an earlier collaboration, they produced the first clear photograph of the sun's surface (1845).

Fizeau and Foucault each followed different procedures for finding the speed of light. Their differences hinged on the question of reaction. Fizeau placed his trust on "toothed wheel" experiments. His experiment required the observer to mark the exact moment when light burst through the dents

---

9. While Fizeau's results centered on 315,000 kilometers per second, Foucault's gave 298,000 kilometers per second. Cornu's results were 298,500 kilometers per second.

of a rapidly spinning toothed-wheel, a moment that depended on the velocity of the wheel, on the observer's perception of that moment, and on his quick reaction to it.[10] Foucault's experimental system avoided all of the reactions associated with timing the appearance and disappearance of signals. To obtain the *"observation of a moving image as a fixed image"* he used a rotating mirror method described by Charles Wheatstone.[11] By reflecting a light beam off a rapidly rotating mirror, Foucault measured the speed of light in terms of its displacement as an angular distance. In this way, a measurement of speed was transformed into a measurement of length. By 1862 his significantly smaller number for the speed of light displaced Fizeau's value, but not all errors had disappeared. Foucault carefully searched for an error in the displaced (although static) images and concluded: "After much research, I ended up by discovering that the cause of error was in the micrometer."[12] His experiments to determine the speed of light were still at the mercy of "errors of observation" because he used a micrometer "comparable in every way to the . . . micrometers used for astronomical observations."[13]

Back to Reaction

A few years after Foucault presented his results, Cornu returned to these experiments by testing the instrumental configurations of Fizeau and Foucault. He first compared Fizeau's toothed-wheel method against Foucault's rotating mirror, and his trust was ultimately placed on the former.[14] In

10. Fizeau's experiments were described in François Arago, "Chapitre XIII: Mesure de la vitesse de la lumière par des observations faites sur la terre à de courtes distances," in *Astronomie populaire, oeuvre posthume*, ed. J.-A. Barral (Paris: Gide, 1857), 422–23.

11. Léon Foucault, "Sur les vitesses relatives de la lumière dans l'air et dans l'eau," *Thèse de doctorat ès sciences physiques* (1853). Reprinted in Léon Foucault, *Mesure de la vitesse de la lumière. Étude optique des surfaces: Mémoires de Léon Foucault*, ed. H. Abraham et al., Les Classiques de la science (Paris: Armand Colin, 1913), 14–44. Citation from p. 26; italics in original. Also reprinted in Léon Foucault, "Sur les vitesses relatives de la lumière dans l'air et dans l'eau," *Annales de chimie et physique* 41 (1854). His concern with fixing movement also appear in Léon Foucault, "Méthode générale pour mesurer la vitesse de la lumière dans l'air et les milieux transparents. Vitesses relatives de la lumière dans l'air et dans l'eau. Projet d'expérience sur la vitesse de propagation du calorique rayonnant," *Comptes rendus des séances de l'Académie des sciences*, 30 (1850): 551–60. Reprinted in *Mesure de la vitesse de la lumière. Étude optique des surfaces: Mémoires de Léon Foucault*, 1–14.

12. The error in the micrometer readings led him to alter his experiments. Instead of measuring the small deviation he adopted a fixed quantity for it and determined the distance separating the sight and the mirror. Léon Foucault, "Détermination expérimentale de la vitesse de la lumière; description des appareils," *Comptes rendus des séances de l'Académie des sciences* 55 (1862): 796. Reprinted in Léon Foucault, *Mesure de la vitesse de la lumière. Étude optique des surfaces: Mémoires de Léon Foucault*, 46–50.

13. Foucault, "Détermination expérimentale de la vitesse de la lumière; description des appareils," 794.

14. In preliminary experiments he installed the toothed-wheel configuration used by Fizeau at the École polytechnique and at the old telegraph tower (rue de Grenelle in Saint Germain). He also tried Foucault's by using a revolving mirror "identical" to Foucault's and also built by Froment. In the latter experimental system he noticed the "petitesse des déviations" that combined with other practical and theoretical problems to convince him of following Fizeau's alternative. Cornu, *Détermination de la vitesse de la lumière*

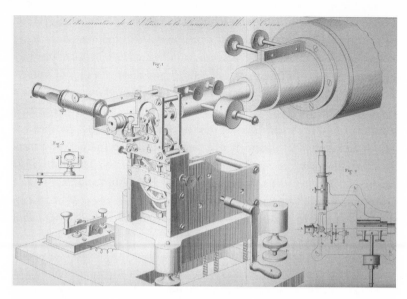

Figure 6.1. Cornu's apparatus showing the changes introduced to the toothed-wheel method by Cornu. Particularly relevant to his goal of eliminating the personal equation was the introduction of a telegraphic key (see lower left corner) to automatically inscribe the moment noted by the observer. From Alfred Cornu, *Détermination de la vitesse de la lumière d'après des expériences exécutées en 1874 entre l'observatoire et Montlhéry, extrait des Annales de l'Observatoire de Paris (Mémoires, tome XIII)* (Paris: Gauthier-Villars, 1876), plate 3.

Fizeau's method he found a clear advantage, "thanks to diverse improvements ... it lends itself to a considerable variety in the methods of measurement, a variety that *permits the elimination of personal errors,* inherent to all methods of observation."[15] Working closely with Fizeau, he altered his experimental system to reduce these personal errors. He followed astronomers and physiologists who introduced chronographic methods of observation by incorporating "an electric key identical to the manipulator of the Morse telegraph" and coupling it with a graphic inscription device. In this way he recorded the observer's reaction to the light signal along with the speed of the rotating toothed wheel.[16]

Between 1871 and 1872 Cornu sent light signals back-and-forth from the small observatory at the École polytechnique to Mont Valérien (separated by 10,310 meters). At the moment of readiness, the observer, with his eye

d'après des expériences exécutées en 1874 entre l'observatoire et Montlhéry, extrait des Annales de l'Observatoire de Paris (Mémoires, tome XIII), 3.

15. Ibid., 2. Italics mine.

16. The use of the graphic method by Cornu was described by Étienne-Jules Marey, *La Méthode graphique dans les sciences expérimentales et principalement en physiologie et en médecine* (Paris: G. Masson, 1878), 480–81.

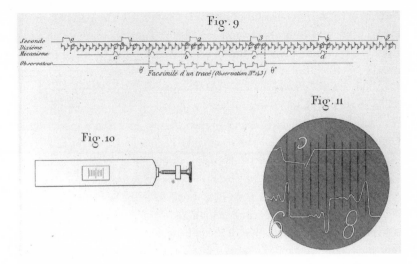

Figure 6.2. The graphic method used by Cornu. This illustration shows the results produced by the observer when signaling with the telegraphic key. From Alfred Cornu, *Détermination de la vitesse de la lumière d'après des expériences exécutées en 1874 entre l'observatoire et Montlhéry, extrait des Annales de l'Observatoire de Paris (Mémoires, tome XIII)* (Paris: Gauthier-Villars, 1876), plate 6.

at the ocular, held a brake regulating the speed of the toothed-wheel in his right hand and a telegraphic key in the left hand. Letting the speed increase, he pressed the key at the time when the signal disappeared, and then, letting the speed decrease, he released the key when it appeared. Despite these changes, this method still encountered some of the same disadvantages of Fizeau's. The observer still encountered an "effect of *surprise*" due to the appearance and disappearance of the light signal, introducing unpredictable errors.[17] To lessen the effects of surprise, Cornu devised a second manner of observing "which one may call *static,* since one waits, almost at rest, for the most favorable moment of signaling."[18] He reduced the effects of reaction time by regulating the appearance of the signal. With practice, the observer could learn to give to the toothed-wheel a periodic change of speed. Once this rhythmic state was established, he could signal during the middle of these periodic changes. Cornu's hope was that, with skill and practice, the appearance and disappearance of the signal would become sufficiently predictable to avoid "surprising" the observer.

17. Cornu, "Détermination nouvelle de la vitesse de la lumière," 156.
18. Ibid.

Urbain Le Verrier, director of the Paris Observatory, supported Cornu's experiments.[19] With his help, he repeated and improved on them, moving from the small observatory at the École polytechnique to Charles Perrault's monumental edifice in Paris.[20] He aimed his instrument even further, to Montlhéry. After a worrisome period when he was not able to see the reflected light, on 25 June 1874 he was finally successful. He shared his success with Fizeau: "I have finally seen the reflected light, the *luminous echo,* as you call it, coming back."[21] Increasing the distance between the reflecting mirrors was an improvement, but Cornu admitted that errors remained.

Some errors were due to the "lost time produced by the persistence of the visual impression on the retina and by the delay in the transmission of signals."[22] Hoping "for a compensation in these sorts of errors" he studied how "the physiological appreciation of the apparition and disappearance of the returning light" affected his results.[23] In a chapter-length "véritable étude des *erreurs personnelles*" he analyzed cerebral and transmission processes that came into play when signaling: "This signal, that defines the speed of the toothed wheel, evidently includes errors specific to the two organs which serve as intermediaries, sight and touch, and an error due to their correlation by the intermediation of the brain. The observed speed is thus tainted by errors due to physiological transmissions, errors that we can call personal and that must necessarily have an influence on the exactitude of the results."[24] Noting that it was "difficult to know *a priori* between which limits the probable error is comprised," Cornu divided it into its essential parts. The first source of error, due to the persistence of retinal impressions, was well known and had been amply investigated. It was, however, entangled with a second source of error that was significantly more complicated. It was the error introduced by "*hesitation,* necessary to the conscious and reflective appreciation of the phenomenon."[25] The third error, due to the time needed to react after the eye received the impression, was also

19. Cornu, *Détermination de la vitesse de la lumière d'après des expériences exécutées en 1874 entre l'observatoire et Montlhéry, extrait des Annales de l'Observatoire de Paris (Mémoires, tome XIII),* 1.

20. Cornu was able to get support and funding from Le Verrier through the intermediary of Fizeau. See letter from Cornu to Fizeau asking for more money for his experiments: "cette note signée de votre main M. Le Verrier ne fera aucune difficulté pour établir deux nouvelles commandes, l'une de 700 francs l'autre de 600." Cornu to Fizeau, Paris, 3 July 1874, Dossier biographique de Cornu, AAS, on p. 4 of 4.

21. Cornu to Fizeau, Paris, 25 June 1874, Dossier biographique de Cornu, AAS, on p. 1 of 4.

22. Cornu, "Détermination nouvelle de la vitesse de la lumière," 156.

23. Ibid., 172.

24. Cornu, *Détermination de la vitesse de la lumière d'après des expériences exécutées en 1874 entre l'observatoire et Montlhéry, extrait des Annales de l'Observatoire de Paris (Mémoires, tome XIII),* 101.

25. Cornu undertook experiments on hesitation anew. His tests on visual and auditory signals revealed that hesitation varied from two- to three-tenths of a second in the former to one-tenth of a second in the latter. Ibid., 102.

difficult to study. Cornu explained how this was the error often studied by astronomers: "It is a cause of error of the genre studied by astronomers, and which constitutes a great part of that which is called *personal error.*"[26] Taking values from electrochronographic observations of meridian transits, Cornu expected it to rarely surpass a tenth of a second.

While physical speed-of-light experiments gained increasing acceptance, they were still haunted by tenth-of-a-second errors. "Unfortunately," Cornu explained, "almost all of these delays are part of the domain of nervous phenomena, that is to say essentially variable, following external circumstances."[27] His conclusion was cautious: "with respect to having eliminated them completely, there is no demonstration which can provide this certitude."[28] The physicist used the French translation of Helmholtz's famous *Handbuch der Physiologischen Optik* for his studies. But by the mid-1870s certain aspects of Helmholtz's optics were starting to become outdated: those which no longer answered to current concerns about reaction time, personal equation, and visual persistence errors.

Inverted Roles

In 1878 Albert A. Michelson, an American physicist who would receive the Nobel Prize in Physics for 1907, turned to the "unfinished business" surrounding speed-of-light experiments.[29] Describing them as expanding on "Cornu's elaborate memoir upon the determination of the velocity of light," he initially spent a mere ten dollars. He perfected the rotating mirror method, which had earlier suffered from problems that prevented the precise measurement of the speed of light. By using a mirror that "could execute 128 turns per second" and increasing the distance to five hundred feet, Michelson augmented the degree of the light's deflection "about twenty times that obtained by Foucault."[30] This change showed a "remarkable

26. Ibid., 102–3. Physiological riddles were finally absent in the fourth "toute mécanique" error, and due to the transmission of the signal form the hand to the graphic inscription device. Cornu, *Détermination de la vitesse de la lumière d'après des expériences exécutées en 1874 entre l'observatoire et Montlhéry, extrait des Annales de l'Observatoire de Paris (Mémoires, tome XIII)*, 103.

27. Ibid.

28. Ibid., 112.

29. For Foucault's and Fizeau's early work on the ether, see Paul Acloque, "Hippolyte Fizeau et le mouvement de la Terre: Une Tentative méconnue," *La Vie des sciences* 1, no. 2 (1984); Pierre Costabel, "L. Foucault et H. Fizeau: Exploitation d'une information nouvelle," *La Vie des sciences* 1, no. 3 (1984). The phrase "unfinished business" is from J. B. Gough, "Armand-Hippolyte-Louis Fizeau," in *Dictionary of Scientific Biography*, ed. Charles Coulston Gillispie (New York: Scribner's, 1970–86), 20.

30. Albert A. Michelson, "Experimental Determination of the Velocity of Light," *Proceedings of the American Association for the Advancement of Science* 28 (1880): 124. He further increased the speed to 250 turns per second and 2,000 feet.

coincidence of the result of these experiments with that obtained by
Cornu, by the method of the toothed wheel."[31] Even before the second transit of Venus approached, Michelson's work was a strong contender among
alternatives for the best determination of the solar parallax.[32]

In light of the success of physical determinations of the speed of light,
defenders of chronophotography suddenly appeared on the defensive. They
stressed the benefits of their methods by underlining their simultaneous
scientific, artistic, and public applications, and by exposing the uselessness
of speed-of-light measurements, asking, with evident irony, what does that
cure? "They talk of the experiments instituted to determine the speed of
light. We would like to know, like Molière's savant, what does that cure?
We would answer that the photoscopic study of movement is an important
aide, some sort of new instrument of analysis of general mechanics, to the
art of the sculptor and the painter, to the physiology of locomotion, and
therefore to hygiene and medicine itself."[33] Astronomers also appeared on
the defensive. Since the publication of the first volume of Laplace's *Mécanique céleste* in 1799, astronomy occupied a privileged place in science. Its
enviable stature was reflected in Laplace's personal choice for a standard of
measurement. He charged scientists to deduce the meter from the earth's
quadrant circumference, a value determined largely through astronomical
observations. Astronomy's position in science, however, started to decline
as the century advanced. Physicists competed with astronomers to tackle
errors of the order of a tenth of a second. By the end of the century, physics—and not astronomy—occupied the top spot. The speed of light, instead
of the length of the meter as a part of the earth's circumference, emerged as
the preferred standard of measurement.

The work of physicists advocating terrestrial speed-of-light measurements went against a century-long tradition where astronomical methods
were used to determine physical constants—not the other way around. For
example, in the seventeenth century, Ole Rømer, the Danish astronomer
who worked at the Paris Observatory, calculated the speed of light from
current astronomical determinations of the solar parallax. Scientists had
long known of the reverse operation: terrestrial measurements of the speed
of light could be used to determine the solar parallax. Yet for years this

31. Ibid., 160. Cornu's work (as reinterpreted by Helmert d'Aix) gave 299,990 kilometers per second for
the velocity of light, while Michelson's gave 299,940 kilometers per second.
32. For an account of Michelson's work and its relation to the solar parallax and the transits of Venus,
see Octave Callandreau, "Histoire abrégée des déterminations de la parallaxe solaire," *Revue scientifique*
2 (1881).
33. Louis Olivier, "Physiologie: La Photographie du mouvement," *Revue scientifique* 30 (1882): 809.

method seemed untrustworthy. Only astronomical measurements would "immediately convince the spirit" of the true parallax value.[34] An important science writer explained how "various methods have been adopted for [determining the solar parallax], but the one which makes use of a transit of Venus has generally been considered to be the most accurate."[35] These hopes appeared to shatter after the results of the transit of Venus were inconclusive. Physicists started to play an increased role with regard to precision measurements. They designed experiments to counter the problems plaguing telegraphic, photographic, and chronophotographic methods used in astronomy. By the mid-1870s, while astronomers and an eager public were still awaiting the results of the transit of Venus expeditions, physicists were busy at work.

Astronomy slowly lost its preeminence, and physicists eventually came to determine most constants—even astronomical ones. In 1875 the attendees of the evening lectures of the British Royal Institution were alerted to the "inverted" roles of physics and astronomy: "Now the progress of science requires an inverse march; the exact value of the velocity of light permits, by the inverted calculus, the computation of the mean distance of the sun or the sun's parallax, that is to say, the same element which is directly given by the transit of Venus."[36] Cornu explained the new order in similar terms: "Today *astronomy reverses those roles* and demands from the progress of Optics the value of this constant."[37] Physical determinations of the speed of light, he argued, were a simpler, cheaper, and surer way of determining the solar parallax. With them, the physicist hoped to contribute to this problem of "capital importance" and "define the absolute dimensions of the solar system."[38]

Le Verrier played an essential role in reversing the hierarchies between astronomy and physics. When he saw that the number he had personally calculated for the solar parallax coincided neatly with the one derived from the speed of light, he called on physicists for help, forcefully backing Cornu. Physical determinations of the solar parallax through speed-of-light

34. George Forbes, *The Transit of Venus*, Nature Series (London: Macmillan, 1874), 17; Charles Delaunay, "Notice sur la distance du Soleil à la Terre, extrait de l'annuaire pour l'an 1866, publié par le Bureau des longitudes," in *Recueil de mémoires, rapports et documents relatifs à l'observation du passage de Vénus sur le Soleil* (Paris: Firmin Didot, 1874), 94.

35. Forbes, *The Transit of Venus*, 17

36. For an account of Cornu's second determinations from the Paris Observatory to Montlhéry (ordered by the Conseil of the Paris Observatory on the proposal of Le Verrier), see Cornu, "New Determinations of the Velocity of Light," published by the Royal Institution of Great Britain on 7 May 1875.

37. Cornu, "Détermination nouvelle de la vitesse de la lumière," 139.

38. Camille Flammarion, "Le Prochain Passage de Vénus et la mesure des distances inaccessibles," *La Nature: Revue des sciences et de leurs applications aux arts et à l'industrie* 2 (1874): 387, Cornu, "Détermination nouvelle de la vitesse de la lumière," 138.

measurements, he insisted, should no longer be considered inferior to astronomical ones. Others agreed. The value of the speed of light, ironically, "found favor among astronomers" at a time when it "was not accepted by most physicists."[39]

Centers of Calculation

The changing status of physics vis-à-vis astronomy was reflected in the expansion of laboratories all over the world dedicated to physics and metrology.[40] The Bureau International de poids et mesures (1874) in France, the National Physical Laboratory (1899–1900) in Britain, the Physikalisch-Technische Reichsanstalt (1887) in Germany, and the National Bureau of Standards (1901) in the United States became prominent metrological institutions.[41] Precision flowed outward from these laboratories. After emerging within European and North American centers, precision science reached the most remote corners of empire.

Physical speed-of-light investigations were widely seen as advantageous precisely because they could be performed in European centers and because they did not depend on the expensive, dangerous, and complicated expeditions that had characterized so much of astronomy during the eighteenth and nineteenth centuries. According to Cornu and his supporters, these experiments were better than astronomical ones because scientists did not have to leave Paris. Parisian speed-of-light experiments represented a stark change from earlier scientific practices in which faraway expeditions were launched to obtain precision constants (for example, during the famous missions of Maupertuis to the Arctic Circle to establish the shape of the earth and of Joseph Delambre and Pierre Méchain to establish its circumference and the size of the meter). Physicists hoped that stable standards could be kept and maintained at centralized laboratories of physics. The role of astronomy as a "natural history" of the heavens, as portrayed in Alexander von Humboldt's sensational *Cosmos,* would slowly wane. Although expeditions would remain important for many other areas of science, precision science would increasingly depend on centers of calculation attached to

39. Cornu, "Détermination nouvelle de la vitesse de la lumière," 139.
40. In the years between 1874 and 1885 laboratories of physics in Britain more than doubled from ten to twenty-four.
41. For the theme of "centers of calculation" see chapter 6 of Bruno Latour, *Science in Action: How to Follow Scientists and Engineers through Society* (Cambridge, Mass.: Harvard University Press, 1987), 216–57. Latour mentions that, according to the National Bureau of Standards, the United States spends 6 percent of its gross national product in maintaining physical constants, that is, three times what is spent on research and development. Ibid., 251.

the world outside via sturdy metrological networks connecting centers to colonial peripheries.

The desire to produce knowledge within European centers can be traced back to Arago, a close friend of the famous explorer Alexander von Humboldt, who had a completely different attitude toward travel. Cornu graced the first page of his famous work on the speed of light with Arago's sage words: "In repeating these observations (on the speed of light) . . . we will one day, without leaving Paris and its environs, find this solar parallax, which since the middle of the last century has led to so long, so far away, so painful, and so expensive voyages."[42] Arago's views about scientific travel can in part be explained by his fate while working in Spain to complete the task of measuring the terrestrial meridian. After the outbreak of the Franco-Spanish war, Arago was imprisoned. Once liberated, he encountered further obstacles when he tried to return to France from embattled Algeria. Le Verrier shared the desire of producing science within Paris. He became the main enemy of scientific expeditions. Finding the transit of Venus expeditions an excuse for "all the functionaries who want to profit from it in order to travel around the world at the expense of the government," he became the chief proponent of physical determinations of the speed of light.[43] Although Cornu presented his results before the 1874 transit, five Navy ships laden with ten kilograms of silver smeared on photographic plates nonetheless sailed off.

Bureau International des Poids et Mesures

In order to bypass the problems faced by traveling astronomers, scientists increasingly found hope in a different set of instruments based on interferometry. Interference effects had been studied centuries earlier by Newton, who investigated the fringes that were formed around thinly layered materials, such as air trapped between two lenses, oil on water, or soap bubbles. In 1851 Fizeau incorporated interferometry into his experiments for measuring the relative speed of light. By passing light into tubes of water

42. François Arago, *Astronomie populaire, oeuvre posthume*, ed. J.-A Barral (Paris: Gide, 1857), 4:418. Cited in Cornu, *Détermination de la vitesse de la lumière d'après des expériences exécutées en 1874 entre l'observatoire et Montlhéry, extrait des Annales de l'Observatoire de Paris (Mémoires, tome XIII)*, 1.

43. LeVerrier to Jurien de la Gravière, Paris, 24 July 1873, AN, F17 3726, Folder: Passage de Vénus (1867–82, particulièrement 1873–75), p. 1. For other accounts commenting on Le Verrier's opposition see "Académie des sciences: Séance du 6 juillet," *Le Moniteur scientifique* 16 (1874); Rodolphe Radau, "Les Applications scientifiques de la photographie," *Revue des deux mondes* 25 (15 February 1878): 885–86; Arthur Schuster, *Biographical Fragments* (London: Macmillan, 1932), 198–201. I thank Simon Schaffer for mentioning this last source to me.

moving in opposite directions and comparing them interferometrically at their exit, he studied how light was dragged by water. In the years after his initial investigations Fizeau applied interferometry to metrology. In 1864, obtaining results that corresponded to a millionth part of a millimeter, he discovered the minute dilations that are now named after him. "A wave length," he wrote "with its series of undulations of an extreme delicacy, can be considered a natural micrometer of the highest perfection, particularly apt for measuring length."[44] By the end of the century his dream would be, for the most part, realized.

Justin-Mirande René Benoît (1844–1922), director of the Bureau international des poids et mesures who "dedicated thirty-seven years of his life, with an indefatigable zeal, to high precision metrology," was most influential in directing research to this promising area.[45] His work on metrology was so important that in the introduction to the 1900 International Congress of Physics, scientists expressed how advances in all areas of science depended on it: "Before one studies the relation between different lengths, it is necessary to first establish what rules one must follow to measure them, that is, to be able to know their true nature."[46]

Benoît sponsored the work of Michelson who later became known for performing experiments to determine the effects of the ether on the velocity of light done in collaboration with Edward Williams Morley. Through their connection to Einstein's theory of relativity, these experiments revolutionized longstanding beliefs about the electrodynamics of moving bodies and the nature of space and time. These precision experiments marked a new configuration between science, action, and reaction that characterized the next century. They were characterized by the elimination of reaction time from measurement techniques (achieved largely through interferometry) and sustained by networked centers of calculation that crowned physics as the ultimate science of precision.

In 1889 Michelson and Morley described a preliminary interferometric experiment for "making a light wave a practical standard of length."[47] While Fizeau's interference effects were only visible across "two or three centimeters," Michelson found that if the light source came from a vacuum tube

44. Armand Fizeau, *Annales de chimie et de physique* 2 (1864).

45. D. Isaachsen, "Introduction historique," in *La Création du Bureau international des poids et mesures et son oeuvre* (Paris: Gauthier-Villars, 1927), 25.

46. Ch.-Éd. Guillaume and Lucien Poincaré, "Avertissement," in *Rapports présentés au congrès international de physique*, ed. Ch.-Éd. Guillaume and Lucien Poincaré (Paris: Gauthier-Villars, 1900), ix.

47. Albert A. Michelson and Edward Williams Morley, "On the Feasability of Establishing a Light-wave as the Ultimate Standard of Length," *American Journal of Science* 38, no. 225 (1889). Another publication on this topic is Albert A. Michelson, "Comparison of the International Meter with the Wave Length of the Light of Cadmium," *Astronomy and Astrophysics* 12 (1893): 590.

illuminated electrically, "interference phenomena could be revealed across distances ten or twelve times as great."[48] A year later, Michelson published a method for measuring the diameter of astronomical objects, another area plagued by personal equation problems.[49] His work caught the attention of Benjamin A. Gould, member of the Comité international des poids et mesures, who secured for him a six-month-long sabbatical from Clark University at Worcester, Massachusetts, where he was professor of physics and invited him to the Bureau in Paris. The initiative was supported by Wilhelm Förster and Adolph Hirsch, two astronomers with firsthand experience with problems of the personal equation.

During the years 1892–93, Michelson worked with Benoît at the Bureau international des poids et mesures to determine standards of length. Michelson's stay in Paris lasted for more than a year, culminating in the construction of a costly apparatus designed to compare the length of the meter against wavelengths from various light sources.

In 1900 Benoît described how the personal equation was one of the main problems preventing scientists from attaining a level of precision beyond the tenth of a second. The personal equation varied according to the "momentary disposition of the observer" and it was "not easy to understand what this phenomenon depends on." In a "history of precision in metrology," where Michelson's work took a central place, Benoît described the dangerous effects of the personal equation on standards of length. Measurements of length standards suffered from a "*personal equation,* analogous to the one in astronomy, and which consists in the propensity that each observer has to appreciate in a certain manner, different from that of others, the position of an axle on a short line or the center of a point, while superposing them with the line of a reticule." He warned of "the influence of differences in appreciations due to different observers" and of how when establishing standards sometimes the "determination becomes uncertain or fixed, even unconsciously from the perspective of the observer . . . which a different observer sees differently." Even the frequent "repetition of observations" or the apparent "accord between numerous measurements . . . does not prove much." It could not prevent the fact that "even under identical conditions, one observer has a tendency to measure *strongly,* while another one measures *weakly.* This divergence in the measurements is, additionally, not

48. Albert A. Michelson, *Détermination expérimentale de la valeur du mètre en longueurs d'ondes lumineuses,* Travaux et mémoires du Bureau international des poids et mesures (Paris: Gauthier-Villars et fils, 1894), 3–4.
49. Albert A. Michelson, "On the Application of Interference Methods to Astronomical Measurements," *Philosophical Magazine* 30 (1890).

Figure 6.3. Michelson's interferometer to measure the meter against light waves. From Albert A. Michelson, *Détermination expérimentale de la valeur du mètre en longueurs d'ondes lumineuses: Travaux et mémoires du Bureau international des poids et mesures* (Paris: Gauthier-Villars et Fils, 1894), plates 1 and 2 at end of book. Other images appear in Ch.-Éd. Guillaume, "L'Oeuvre du Bureau international des poids et mesures," in *La Création du Bureau international des poids et mesures,* edited by Ch.-Éd. Guillaume, 35–258 (Paris: Gauthier-Villars et Cie, 1927), p. 205.

absolutely constant: it can vary according to the particular conditions of observation, and also, with time, according to the momentary disposition of the observer." Personal equation errors coupling with other types of errors made fixing a standard nearly impossible. According to Benoît, all matter seemed endowed with movement and even life: "We recognize that matter, even if it first appears to us immobile and inert, is in reality the seat of an interior activity constantly in play, and one can reasonably say that it is endowed with some sort of life."[50]

Michelson's interferometric experiments solved some of the measurement problems described by Benoît, restoring to matter its immobility and inertness. By not requiring an observer's reaction to a signal, Michelson avoided tenth-of-a-second reaction time errors. Yet Michelson was forced to

50. Benoît was director of the Bureau from 1889 to 1915. J.-René Benoît, "De la précision dans la détermination des longueurs en métrologie," in *Rapports présentés au congrès international de physique réuni à Paris en 1900,* ed. Ch.-Éd. Guillaume and L. Poincaré (Paris: Gauthier-Villars, 1900), 61, 62.

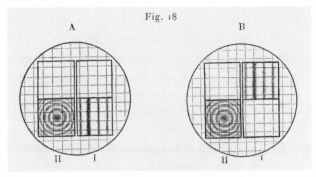

Figure 6.4. *A* and *B*, image at the ocular of Michelson's interferometer. An observer had to focus the interference rings at the lower left quadrant to obtain a minimum of light at the center. In this process, a personal equation appeared. From Albert A. Michelson, *Détermination expérimentale de la valeur du mètre en longueurs d'ondes lumineuses: Travaux et mémoires du Bureau international des poids et mesures* (Paris: Gauthier-Villars et fils, 1894), p. 47.

study personal equation errors that appeared in calibrating his instrument. When an observer saw interference fringes at the ocular, these needed to be focused in order to obtain a minimum of illumination at the center of the rings. In this operation, a personal equation came into play: "It exists almost inevitably in this observation a type of *personal equation,* which is unique to each observer, and that varies perhaps, for the same observer from one time to the next, and according to the size and the clarity of the observed rings."[51] Yet, apart from this initial operation, he was able to eliminate the personal equation in almost every other way by transforming the most important task of the observer into a process of counting wavelengths in whole numbers. These numbers were large, in the thousands. Their precise determination could be checked by having multiple observers count them. When Benoît and Michelson each counted them, Michelson was led to the conclusion that "the almost absolute accord, throughout all the determinations, of the numbers as deducted from a series of observations completely independent from each other, *and made by different observers,* constitutes a new confirmation of the exactitude of the whole process."[52] The exactitude of his instrument not only resided in a coherence in the results obtained by a single observer, but—most importantly—by the agreement in the observations of two or more experimenters.

51. Michelson, *Détermination expérimentale de la valeur du mètre en longueurs d'ondes lumineuses,* 79.
52. Ibid., 80. Italics mine.

The Ether

In the course of investigating the solar parallax and the speed of light, scientists became increasingly preoccupied with a third element: the ether, an elusive substance believed to pervade all space that served as a lynchpin for classical mechanics. To arrive at a correct calculation for the solar parallax from terrestrial speed-of-light measurements, scientists needed to determine the effect of the ether on the velocity of light—that is, they needed to account for the resistance, called the ether-drift, encountered by light as it moved through space.[53] Physical determinations of the speed of light eliminated the effect of the ether because the back-and-forth trajectories of the ray of light cancelled its effect. This characteristic, noted by Cornu at the prompting of the astronomer Yvon-Villarceau and later brought up by the physicist and founder of electromagnetism James Clerk Maxwell, proved to many scientists the difficulties of obtaining astronomical constants through physical experiments on the speed of light.[54] They also showed how speed-of-light experiments were intertwined with ether research. For Cornu, speed-of-light measurements "contribute just as well to important questions about the constitution of the hypothetical milieu, seat of luminous movement, the *ether*."[55] For the astronomer Simon Newcomb, the "more generally recognized importance" of speed-of-light measurements resided in their relation to determining the distance of the sun from the earth, but equally important were questions about "the qualities of the luminiferous ether and the relation between light, electricity, and other visible forces."[56]

53. The two methods (one based the observation of Jupiter's satellites and the other on the phenomenon of annual aberration) used for undertaking the conversion between astronomical values and physical values, depended on the behavior of light as it moved through the ether. The first method depended on the realization that when one of Jupiter's satellites passes into the shadow of the planet it is seen 480 seconds earlier when Earth is on the same side of the sun as Jupiter than when Earth is on the opposite side. Since the distance between Jupiter and Earth is shorter by a whole diameter of Earth's orbit when Earth is on the same side of the sun as Jupiter, then, assuming that light travels at around 298,500 kilometers per second, the distance of Earth to the sun would be about 298,500 kilometers per second multiplied by 480 seconds: 143,280,000 kilometers. This was the method used by Rømer for determining the speed of light from current values of the solar parallax in 1676. The second method depended on the realization that the length of the Earth's circumference could be found by multiplying the velocity of the Earth's orbit around the sun by the number of days in a complete cycle (365); once the circumference of the Earth's orbit was known, its distance from the sun could be calculated. The velocity of Earth could be found through the phenomenon of stellar aberration, which gave Earth a speed of 1/10,000th times the speed of light. This method, which was used by Cornu, depended on the concept of a stationary ether as developed by Augustin Fresnel and George Stokes. When Michelson undertook his experiments on the "Relative Motion of the Earth and the Luminiferous Ether" and found no effect from the ether, he concluded that his experiment contradicted "the explanation of the phenomenon of aberration, which has been hitherto generally accepted, and which presupposes that the earth moves through the ether, the latter remaining at rest."
54. See letter from James Clerk Maxwell to David Todd Peck, James Clerk Maxwell, *Nature* 21 (1880).
55. Cornu, "Détermination nouvelle de la vitesse de la lumière," 139.
56. Simon Newcomb, "On a Proposed Modification of Foucault's Method of Measuring the Velocity of Light," communicated to the National Academy of Science, April 1878, in Simon Newcomb Papers, Li-

Due to the problems of reaction facing astronomy, the question of the speed of light and its relation to the ether moved to center stage, a process that started with the work of Foucault and Fizeau and was continued by others, such as Michelson. The competition between physicists and astronomers intensified the desire to understand how the ether affected the velocity of light. Michelson's ether experiments differed in essential ways from speed-of-light measurements. Yet both of these experiments were directly related to the problem of measurement errors.[57] In fact, his Nobel Prize was awarded in recognition for the interferometric experiments considered to be some of the most precise ever—not for his work on the ether. The reasons why they were so precise depended on certain assumptions about the tenth of a second and its relation to personal errors.

The Jacobin's Ideal

Michelson's work with Benoît and "in collaboration with the personnel of the Bureau" was central to the goal of eliminating sensorimotor errors and a direct response to competing photographic and cinematographic methods.[58] By the turn of the century they were physicists' best hope for solving the problem of standards. According to Benoît, "The work of Professor Michelson, the perfection he introduced in interferometric methods, the discovery of new sources of light more monochromatic than those known before, have given the first solution to the problem."[59] With these changes, measurement errors could be reduced to a millionth of a meter. Benoît boasted how "in the space of two centuries, the precision of measurement has augmented approximately in the order of 1 to 1000."[60] While still in Paris, they famously measured the meter against the wavelength of cadmium light.[61]

Charles Édouard Guillaume, Benoît's successor as director of the Bureau, expressed the importance of Michelson's work for metrology: "The classic research of Michelson and Morley . . . has placed in the hands of physicists rays which can be standards of length and units, at the same time as

---

brary of Congress, 57. Cited in Richard Staley, "Michelson and the Observatory," in *The Heavens on Earth: Observatory Techniques in Nineteenth-Century Science*, ed. David Aubin, Charlotte Bigg, and H. Otto Sibum (Durham, N.C.: Duke University Press, forthcoming).

57. For Michelson's work in the context of the astronomical concerns pertaining to the solar parallax, see Staley, "Michelson and the Observatory."

58. Benoît, "De la précision dans la détermination des longueurs en métrologie," 61, 62.

59. Ibid.

60. Ibid.

61. For a description of the work of Michelson with Benoît, see Jules Andrade, *Le Mouvement: Mesures de l'étendue et mesures du temps* (Paris: Félix Alcan), 264–66.

measurement instruments."[62] In 1904 Michelson's standard was adopted at the International Solar Union meeting at the St. Louis World's Fair. A few years later this standard received further confirmation with the work of Alfred Pérot and Charles Fabry who built another interferometer to determine the length of the standard meter.[63] This standard finally trumped the eighteenth-century dream of comparing the meter against the circumference of the earth.

Describing these new interferometric techniques, where the observers' task could be reduced to counting in whole numbers, Benoît explained how in the fin-de-siècle scientists were very close to reaching the revolutionary ideals of standards once supported by the Jacobins. Referring to experiments comparing the metric unit to a wavelength produced under fixed conditions, he explained: "In this way, it is interesting to note, how the primitive conception of relating the unit of length to a natural, constant reference (or one that is reproducible at will in an identical form) which was considered for a longtime chimerical, has returned in the fin-de-siècle." Benoît acknowledged that basing standards of measurements on wavelengths was perhaps not as grandiose as the Jacobin's attempt to base the meter on the circumference of the earth, but this idea "in the end is not so different."[64] The success of new metrological laboratories, such as the Bureau international des poids et measures led by Benoît, depended on these solutions to the personal equation.

Destabilizing Moments: Mach and Duhem

During the second half of the nineteenth century, the exact moment and place of separation between physical and psychical domains was a topic of intense debate in physics. But by the first decade of the twentieth century, these questions had mostly disappeared from the discipline. In 1909 the physicist Max Planck, one of the leading figures in theoretical physics, forcefully claimed that the greatest scientific accomplishment of the century had been the "real obliteration of personality."[65]

The triumph announced by Planck was nonetheless controversial at the time. Planck fought against a small group of physicists that allied them-

62. Ch.-Éd. Guillaume, "L'Oeuvre du Bureau international des poids et mesures," in *La Création du Bureau international des poids et mesures et son oeuvre*, ed. Ch.-Éd. Guillaume (Paris: Gauthier-Villars, 1927), 198.
63. Benoît, "De la précision dans la détermination des longueurs en métrologie," 69, 70. In 1967 the meter was redefined as 1,650,763.73 times the wavelength of the orange-red light emitted by 86Kr, a natural krypton isotope.
64. Ibid., 69.
65. Max Planck, *Eight Lectures on Theoretical Physics*, trans. A. P. Willis (Mineola, N.Y.: Dover, 1998).

selves with the Viennese physicist and philosopher Ernst Mach. In the 1870s Mach instructed his assistant to study the personal equation in the sense of vision, concluding that "a parallel of this so-called *personal equation* of astronomers, having its ground in analogous facts, is also observed in the same sense province."[66] The personal equation could cause "an optical impression which arises physically later" to "appear to occur earlier," such as when a "surgeon in bleeding, first sees the blood burst forth and afterwards his lancet enter."[67] As in those illusions which occurred in time, Mach observed analogous anomalies in the sensation of space.[68]

Some of Mach's references to the personal equation, appearing in his seminal *Beiträge zur Analyse der Empfindungen* (1886), were used as part of a series of examples illustrating the need to fight against the "unwonted prominence" of physics in "modern times." Instead of following the physicalist tendency that Mach felt dominated scientific inquiry, he instead strove to point out the impossibility of finding a surefire distinction between "physical" and "psychical" realms.[69] This separation was artificial, he believed, emerging from widespread "habitual stereotyped methods of observation."[70] Mach's approach was radical: he recommended that scientists completely erase this "dividing-line."[71] He advocated broad institutional changes: recombining the disciplines of physics, physiology, and philosophy and curtailing the élan of theoretical physics that was distancing science even further from sensations. For Mach, the example of the personal equation was used to stress the dependency of physics on a *"larger* collective body of knowledge," one which included not only physiology but also philosophy.

Few followed Mach's lead. Einstein, who was initially seduced by his ideas, ultimately found them infertile. Vladimir Lenin, exiled in Siberia, wrote a scathing diatribe against him preempting the threat that his philosophy could pose to the materialism he was developing.[72] Most

---

66. Ernst Mach, *Contributions to the Analysis of the Sensations*, trans. C. M. Williams (La Salle, Ill.: Open Court, 1897 [1886]), 112. Italics mine. For his assistant's "series of experiments which he carried out at my desire," see Vinko Dvorak, "Ueber Analoga der persönlichen Differenz zwischen beiden Augen und den Netzhautstellen desselben Auges," *Sitzber. d. königl. Böhm. Gesellschaft der Wissenschaften* (Math.-naturw. Classe), 8 March 1872. Mach, *Contributions to the Analysis of the Sensations*, 112 n. 2.

67. Mach, *Contributions to the Analysis of the Sensations*, 112.

68. Ibid., 61.

69. For Mach's work on the psychophysiology of time with some references to the personal equation, see Claude Debru, "Ernst Mach et la psychophysiologie du temps," *Philosophia Scientiae* 7, no. 2 (2003): 72, 84, 87. Mach, *Contributions to the Analysis of the Sensations*, 1, 179–80.

70. Mach, *Contributions to the Analysis of the Sensations*, 14.

71. "If, for the higher purposes of science we erase this dividing-line [between the physical and the psychical], and consider all connections as equivalent, new paths of investigation cannot fail to be opened up." Ibid., 152.

72. Lenin denounced Mach in chapter 3 of *Materialism and Empirio-Criticism* (1909).

scientists—and an occasional revolutionary such as Lenin—worked on exactly the opposite the project, on strengthening the "dividing-line" between the physical and the psychical.

The controversial scientist and philosopher Pierre Duhem allied his philosophy with that of Mach. In his famous *Aim and Structure of Scientific Theory* (1905), Duhem also cited the personal equation, deriving other conclusions. He criticized the concept of a "crucial experiment" (often associated with Francis Bacon), where experiments were decisive in settling scientific controversies. In an often-cited and controversial phrase, Duhem claimed that "a crucial experiment is impossible in physics."[73] He also offered a critical view of the pride and joy of experimental scientists: speed-of-light experiments. While he first took issue with Fizeau and Foucault's experiments, he later expanded his criticisms to the concept of experimentation itself. Most of Duhem's contemporaries, and Arago in particular, believed that since Fizeau and Foucault's experiments demonstrated that light traveled more slowly in water than in air, its nature was wavelike. Yet Duhem claimed that "it would be difficult for us to take such a decisive stand."[74] Instead, he urged scientists and philosophers to think about the scientific method more carefully, more critically, philosophically, and historically.

Duhem famously argued that when different scientific theories were compared against each other scientists could not determine which of them was true with a single experiment. Scientific theories were based on many different hypotheses, and numerous assumptions sustained them. In order for scientists to decide among them, "it would be necessary to enumerate completely the various hypotheses which may cover a determinate group of phenomena; but the physicist is never sure he has exhausted all the imaginable assumptions."

In chapters leading to his famous statement on experimentation, Duhem took pains to enumerate the numerous hypotheses and assumptions that came into play in scientific practices. Before "we can accept offhand the judgments in which [the experimenter] states the results of his experiments," scientists had to consider a number of additional factors. Some of these had to do with the theories employed by the experimenter; others with evaluating the relation between the experimenter's "real apparatus" and the "ideal and schematic one on which he did his reasoning." Yet even after "assuming that the experimenter has employed, in order to interpret his observations, theories which we accept with him, that he has correctly

73. Pierre Duhem, *The Aim and Structure of Physical Theory*, trans. Philip P. Wiener (New York: Atheneum, 1962), 188.
74. Ibid., 190.

applied in the course of this interpretation the rules that these theories pre-
scribe, that he has minutely studied and described the apparatus he used,
and that he has eliminated the causes of systematic error or corrected their
effects—that would still not be sufficient reason to accept the result of his
experiment." The task was indeed even more complicated—on various occasions Duhem described it as "infinite." Even the simplest of observations
had to be meticulously checked. To obtain information on the "exactness
of an experiment" scientists judged "the acuteness of the observer's sense,"
and this involved the riddle of the personal equation: "Astronomers try
to determine this information in the mathematical form of the *personal
equation,* but this equation partakes very little of the serene constancy of
geometry, for it is at the mercy of a splitting headache or painful indiges-
tion."[75] The personal equation was one reason proving that the "certainty
of physical experiment remains constantly subordinated to the confidence
inspired by a whole group of theories."[76]

   While basing part of his evidence on the effects of the personal equation, Duhem repeatedly stressed the theoretical biases of *all* scientific observations. These biases were evident *every time* scientists tried to eliminate
errors from their experimental systems. Even looking through a magnify-
ing glass, insisted Duhem, required treating it theoretically, disregarding
the rainbow colored circles that appeared around the enlarged object. The
more complicated the instruments became, the more theoretical assumptions came into play. But even the simplest observation, the simple act of
seeing in science, already carried a marked theoretical bias. Duhem used
the example of the personal equation as a call for rethinking the certainty
of physical experiments, and to rehabilitate the importance of "everyday
testimony" arrived at "by non-scientific methods through mere seeing or
touching."

   Both Duhem and Mach increasingly distanced themselves from main-
stream physics. They remained skeptical of atomic theory and theoretical
physics, repudiating transformations that increasingly dominated the field.
Posterity would remember them as retrograde scientists that nonetheless
made important contributions to the philosophy of science. Mach became
an inspiration to the logical positivists of the Vienna Circle, initially named
the Ernst Mach Verein when founded in 1928. Duhem was revived by the
American philosopher Willard Van Orman Quine, who cited him in the famous "Two Dogmas of Empiricism" (1951) essay. The idea that experiments

75. Ibid., 162. Italics mine.
76. Ibid., 163.

could not unambiguously refute scientific theories, but rather that observation and theory were complexly interrelated became known as the Duhem-Quine thesis, a staple tool for sociologists of science in the twentieth century, who continued to look for marks of society in the most impersonal elements of science.[77]

Solutions to tenth-of-a-second measurement errors by physicists stabilized the discipline to an unprecedented degree, providing a new authority to science and quelling longstanding debates about the relation between physical and psychical domains. These questions, however, did not entirely disappear. While once pertinent in science, they increasingly came to interest philosophers of science. While once directly related to tenth-of-a-second errors, they slowly shed this all too human lineage to become much more.

77. Sandra G. Harding, ed., *Can Theories be Refuted? Essays on the Duhem-Quine Thesis* (Dordrecht: Reidel, 1976).

CHAPTER 7

REACTING TO RELATIVITY

The unraveling of film corresponds to a certain duration of our interior life—
to that one and to no other. The film which unravels is therefore truthfully
attached to a conscience that endures, and that regulates its movement.

HENRI BERGSON, philosopher

During the interwar period, tenth-of-a-second moments were invoked by
scientists when discussing the most revolutionary achievement of their
time: Einstein's theory of relativity. At the time of its inception, important
scientists thought of relativity theory in terms of what they knew about the
tenth of a second.

In a 1921 volume of the prestigious *Comptes rendus* of the Académie
des sciences filled with new essays about the redefinition of time proposed
by Einstein, scientists explored the theory's relation to the tenth of a second. Charles Richet, one of the prime investigators of the personal equation and reaction time, examined the new theory in light of his decades-old
work on "the unit of psychological time."[1] Richet summarized work that he

---

1. Charles Richet, "L'Unité psychologique du temps," *Comptes rendus des séances de l'Académie des sciences*
173, no. 25 (1921).

had begun back in the 1870s. From years of research he had come to the conclusion that "nobody could make more than 10 voluntary movements per second."[2] An "elementary volition" lasted roughly 1/12.5 seconds and if "visual (or auditory) impressions succeed each other at speeds higher than 12 per second . . . the fusion of images begins." He associated these effects with many others, such as light and sound vibrations and the duration of an electric spark. They were "indissolubly tied to the function of our organs and the workings of our conscience," and consequently "an integral part of our humanity." Not even traveling in outer space at the speed of light (as recounted in the fantastical twin paradox story) could, from the perspective of the person traveling, alter these effects.[3]

Richet's views about the tenth of a second's essential connection to our notion of time were widely shared. Both members of the scientific community and the general public tried to figure out how Einstein's theory fit with this research, and the "press referred frequently" to these topics.[4] A writer in the widely read newspaper *L'Écho de Paris* confronted these questions and reframed Einstein's theory in terms of Richet's research: "Following Charles Richet" the "physiological unit of human time" is "1/12 of a second." This unit, according to the writer, arose from "a long adaptation to our milieu" and thus "could not vary brusquely in an astronomical milieu."[5] The author concluded that talk of time expanding and contracting in physics and astronomy would remain irrelevant for human concerns and thus its "revolutionary" implications were profoundly limited.

Some of these questions surfaced in a different form in a famous confrontation between Henri Bergson and Albert Einstein in which the perception of short moments of time played an important role. In contrast to Richet and other scientists, Bergson did not cite the tenth of a second directly, yet he repeatedly wrote about the time needed for images to fuse

---

2. Ibid., 1314.

3. "Even if we are sent to space, by some more prodigious form of transportation, time may well seem for the stationary observer much longer or shorter, but that would matter little, because, from our human point of view, the succession of physiological phenomena would be the same, and any chronometers transported with us would mark the same relative durations." Richet was less concerned with time expanding or contracting for one traveler in comparison to another one, but rather, he strived to show that through the individual perspective of both travelers nothing would change, not even their relation to the tenth of a second. Ibid., 1317.

4. Michel Biezunski, *Einstein à Paris: Le Temps n'est plus* . . . (Vincennes: Presses Universitaires de Vincennes, 1991), 105.

5. Montluys, *L'Écho de Paris*, 29 March 1922, Document 4.12 in Michel Biezunski, *Einstein à Paris: Le Temps n'est plus* . . . (Vincennes: Presses Universitaires de Vincennes, 1991), 106. Another article from *L'Écho de Paris* that referred to Charles Richet was reprinted as document 4.11 in ibid., 106.

in cinematography.[6] The time required for them to fuse proved to the philosopher that something about the passage of time was intimately attached to human consciousness: "The unraveling of film corresponds to a certain duration of our interior life—to that one and to no other. The film which unravels is therefore truthfully attached to a conscience that endures, and that regulates its movement."[7] Bergson was not concerned with setting an actual value for this speed. What was important for him was the simple realization that "in fact, this speed is determined." It proved to him the limitations of Einstein's concept of time. Arguments that had commenced as deliberations on Einstein's theory of relativity in light of an individual's relation to tenth of a second, grew, in Bergson, to a complete philosophy of time, science, technology, and consciousness. From the early 1920s, when Bergson first turned to relativity, to his final work published in 1934, the philosopher thought about time in relation to these moments.

A new understanding of short moments of time was connected to the rise of physics and its establishment as the privileged science of modernity. It was also related to the concomitant decline of philosophy and its prestige vis-à-vis science. Einstein's way of thinking of short moments of time reflected how they were handled in the sciences of his era. When he thought about them he saw no mystery. He considered two ways of understanding them. The first was in terms of physics, as they were revealed by numerous laboratory instruments. The second was psychologically, as they were imperfectly captured because of our sensorial limitations. For Bergson, in contrast, Einstein's views were excessively simple and ignored evident complexities behind these short moments. Yes, instruments had demystified them to some degree. However, we still needed to think about how instruments achieved this. For Bergson, a philosophy of science and technology was just as important as science and technology themselves.

---

6. Bergson's cinematographic critique expanded even further in the 1970s, growing to encompass historical discourse. Michel Foucault, in the *Archaeology of Knowledge*, compared the limitations of discourse to the limitations of the cinematographic method. "Discourse," he explained "is snatched from the law of development and established in a discontinuous atemporality. It is immobilized in fragments, precarious splinters of eternity. But there is nothing one can do about it: several eternities succeeding one another, a play of fixed images disappearing in turn, do not constitute either movement, time or history." Cinematography was used by Foucault as a model for understanding the laws of discourse and establishing the rules for its development. By reference to it, he proposed ways of thinking outside of its limits. Michel Foucault, *The Archaeology of Knowledge*, trans. A. M. Sheridan Smith (New York: Pantheon, 1972), 166. Cited in Friedrich Kittler, *Gramophone, Film, Typewriter*, ed. Timothy Lenoir and Hans Ulrich Gumbrecht, Writing Science (Stanford, Calif.: Stanford University Press), 117.

7. Henri Bergson, "La Pensée et le mouvant," in *Oeuvres* (Paris: Presses Universitaires de France, 1991 [1934]), 1262.

Einstein against Bergson

In 1922 the scientist and the philosopher met in Paris to discuss the effects of relativity on time and simultaneity.[8] At that time, Einstein was a rising star in science. Arthur Eddington's 1919 eclipse expedition had brought him international fame.[9] Partly because of his vocal pacifist and antinationalist stance, Einstein was the one German-born scientist to whom many members of the international community gladly turned. He received the Nobel Prize for Physics in 1921. His visit was highly symbolic for the two countries.[10] After receiving three invitations in 1922 (the last one from the Collège de France) Einstein declined all of them. He had, however, second thoughts about the last one. These doubts intensified after a conversation with the foreign minister, Walther Rathenau, who worked to improve relations between the two nations and who urged him to attend. Shortly thereafter he withdrew his refusal, notified the Prussian Academy of Sciences, and prepared his trip.[11]

Bergson's vita was similarly brilliant. At the Lycée Condorcet he had obtained prizes in English, Latin, Greek, and philosophy. He was acclaimed for his mathematical work, receiving a national prize and publishing in the *Annales de mathématiques*. He published two theses—one a highly specialized dissertation on Aristotelian philosophy and another (*Time and Free Will*) that would go through countless editions. In 1898 he became a professor at the École normale; in 1900, he moved to the Collège de France. In 1907, his fifth book, *Creative Evolution,* brought him universal fame. His lectures at the Collège de France were so crowded with *tout Paris* that his students could not find seats. (It was rumored that socialites sent their servants ahead of time to reserve them.) During his reception at the Académie Française he received so many flowers and so much applause that underneath the clamor he was heard protesting: "Mais . . . je ne suis pas une danseuse!" Even the Paris Opera, it was evident, was not spacious enough for him.[12] This universal fame followed him until 1922, when he published *Duration*

8. This chapter is an elaboration of Jimena Canales, "Einstein, Bergson, and the Experiment That Failed: Intellectual Cooperation at the League of Nations," *Modern Language Notes* 120, no. 5 (2005).

9. Matthew Stanley, "An Expedition to Heal the Wounds of War: 1919 Eclipse and Eddington as Quaker Adventurer," *Isis* 94 (2003).

10. On this topic see Biezunski, *Einstein à Paris: Le Temps n'est plus . . . .*

11. Walter Rathenau was assassinated on 24 June 1922. For details on Einstein's decision to go to Paris, see Otto Nathan and Heinz Norden, eds., *Einstein on Peace* (New York: Avenel, 1960), 42–54.

12. These biographical details are mostly from P. A. Y. Gunter, ed., *Bergson and the Evolution of Physics* (Knoxville: University of Tennessee Press, 1969), 3–42.

and Simultaneity, a book that he described in its preface as a "confrontation" against Einsteinian interpretations of time and that was in press at
the time of their debate.

On numerous occasions Bergson took pains to stress that he held no
grudge against Einstein as an individual and had no qualms about the physical nature of Einstein's theory. Bergson, who was also Jewish, thus differentiated his position from the racist and nationalist attacks that Einstein
encountered in Germany. He objected only to certain *philosophical* exten-
sions of relativity. Uses that, he claimed, arose from a confusion prevalent
"in those who transform this physics, *telle quelle,* into philosophy."[13] Bergson reacted against a perceived encroachment of physics on philosophy.[14]
In the preface to the first edition of *Duration and Simultaneity* he stated his
motivation for pursuing a confrontation. It arose from a "devoir," which
hinged on defending the place of philosophy vis-à-vis science: "*The idea
that science and philosophy are different disciplines meant to complement
each other . . .* arouses the desire and also *imposes on us the duty* to proceed
to a confrontation."[15]

During their face-to-face meeting, time was short, and Bergson was forced
to summarize parts of his argument.[16] In noting the advanced time of day
that evening, he made a philosophical point: "Because no one can speak about
time without noticing the hour, and that it is late now, I will limit myself to
summarizing one or two points." The philosopher insisted on thinking about
time in relation to consciousness, individuals, and life, but Einstein disagreed.
For Einstein, time and simultaneity were "independent of individuals."

In the course of their conversation, Einstein separated time into three
components: "physical time," "psychological time," and "philosophical
time." These divisions were fewer than the ones in earlier debates about
the standardization of time, when scientists spoke of the different times
used by astronomers, physicists, physiologists, engineers, electricians, railroad companies, industrialists, and the public. But for Einstein, most of
these definitions could be subsumed by these three broad divisions. Con-
troversially, Einstein argued that the third category simply "did not exist"
in reality. Time should be understood either psychologically or physically,

---

13. Henri Bergson, "Préface," in *Mélanges* (Paris: Presses Universitaires de France, 1972 [1923]), 60.
14. For Bergson's view of the relation of science to philosophy in comparison with Léon Brunschvicg,
see Frédéric Worms, "Entre critique et métaphysique: La Science chez Bergson et Brunschvicg," in *Les
Philosophes et la science,* ed. Pierre Wagner, Folio essais (Paris: Gallimard, 2002).
15. Bergson, "Préface," 59. Italics mine.
16. For details on the date of its appearance, see "Notes des éditeurs," in Henri Bergson, *Mélanges* (Paris:
Presses Universitaires de France, 1972), 1600.

but not philosophically: "the time of the philosophers does not exist, there remains only a psychological time that differs from the physicist's."[17] Einstein argued that philosophical time was based on overlapping notions of psychological time and physical ones: "The time of the philosopher, I believe, is a psychological and physical time at the same time."[18] But since psychological notions of time were "nothing more than mental constructs, logical entities," no real overlap existed between psychological conceptions and physical conceptions of time.[19] He, therefore, did not see a role for philosophy in matters of time.

The physicist's view of philosophical time was both radical and provocative—especially since he was conversing with *the* philosopher widely considered an authority on time and especially after being hosted by the prestigious Société française de philosophie (while other scientific societies, including the Société française de physique, simply refused to open their doors to him).[20]

Henri Piéron, an experimental psychologist who instituted reaction time measurements in the laboratory of the École des hautes études, intervened in the conversation and brought up the example of the personal equation.[21] Piéron belonged to a new generation of researchers (including Edouard Toulouse, Henri Laugier, and Jean-Maurice Lahy) who succeeded the pioneers who had worked with Théodule Ribot (Henri Beaunis, Alfred Binet, and Pierre Janet). Their work flourished in laboratories of experimental psychology and industrial laboratories dedicated to increasing efficiency. Their success was comparable to developments taking place elsewhere in Europe and in the United States, where students of Wilhelm Wundt (Hall, Warren, Titchener, Cattell, and Scripture) joined others (Angell, Woodward, Sanford, and Münsterberg) in institutionalizing reaction time ex-

---

17. Henri Bergson, "Discussion avec Einstein," in *Mélanges* (Paris: Presses Universitaires de France, 1972), 1344.
18. Ibid.
19. Ibid.
20. Even more ironically, the president of the Société française de physique attended all of Einstein's conferences at the Collège de France. On those who accepted or rejected Einstein, see Charles Nordmann, "Einstein à Paris," *Revue des deux mondes* 8 (1922): 926–37.
21. Henri Piéron succeeded Alfred Binet in the laboratory. Bergson defended the chair of experimental psychology in "Assemblée de Professeurs, Chaire de psychologie expérimentale et comparée," Paris, 10 November 1901, 2 pp., AN, F17-13556, Folder: Chaire de psychologie expérimentale et comparée. Ancienne chaire de droit de la nature et de gens. He mentioned Wilhelm Wundt as the main advocate of laboratory-oriented psychology. Bergson backed Pierre Janet. For Bergson's role in the 1902 elections for the chair at the Collège de France, see Serge Nicolas and Ludovic Ferrand, "Pierre Janet au Collège de France," *Psychologie et Histoire* 1 (2000). For the context of psychology at the time, see Edouard Toulouse, Vaschide and Piéron, *Technique de psychologie expérimentale* (Paris: Octave Doin et fils, 1911).

periments in academic laboratories and in developing its numerous social applications.

For Piéron, the debate between Einstein and Bergson could be clarified by understanding how the personal equation affected determinations of time. This example could illuminate Bergson's argument ("This is a concrete example of the type pointed out by Bergson . . ."). The psychologist proceeded to explain how it arose in astronomy: "For a long time now, astronomers have recognized the impossibility of using a psychological definition of simultaneity in order to determine a physical simultaneity, when they have tried to determine the position of a star across a telescope's reticule at the signaling of a clock." If the enormous speed of light had caused this realization to arrive slowly for physicists, the slow speed of nerve transmission had made it evident a long time ago for physiologists, psychologists, and astronomers. They had long known that perceptions of simultaneity differed from physical simultaneity. Legend had it that most scientists had learned this lesson as early as 1796.

The psychologist argued that the problem had been largely solved by physicists and astronomers when they employed graphic instruments to translate a temporal event into a spatial one. By pressing a telegraphic key at the moment a star crossed the cross wires of their sighting instruments and recording it automatically on a strip of paper, they significantly reduced their personal equations: "It is by the coincidence or by the lack of coincidence of traces left by signaling machines on a surface animated by a more or less rapid movement that we judge physical simultaneity, while taking into consideration all relevant corrections." According to Piéron, these new technologies and laboratory practices explained why Bergson and Einstein were talking at cross purposes.

Piéron argued that Einstein and Bergson were really discussing two different notions of time: "physical time" (Einstein's), which was recorded by these instruments and which "demands a spatial translation," and "philosophical time" (Bergson's), which remained temporal and could never be translated spatially. He concluded: "It is in that way that Bergsonian duration appears to me as having to remain different from physical time in general and in particular to Einsteinian time."

Einstein's confidence in stressing the independence of physical time from individuals' actual assessment of time arose in the context of the new technologies described by Piéron. The physicist belonged to a younger generation than Bergson, for whom new techniques for eliminating the personal equation and reaction time errors from time determinations were commonplace. He was even younger than Max Planck, who decades earlier had already framed the main achievements of modern physics as resulting in the "real obliteration of personality."[22]

Reaction Time Research

Bergson agreed, for the most part, with these interpretations, including Piéron's. But in contrast to the psychologist, the philosopher did not simply want to counter a physical definition of time with a psychological one. Bergson's central mission during their encounter consisted in connecting all of these concepts of time back to philosophy. His philosophy proposed a new way of thinking about technology, including the new instruments and techniques used by astronomers and physicists to determine time.

Bergson's philosophy of technology was most forcefully developed by reference to the cinematographic camera in *Creative Evolution* (1907), but it started to appear much earlier. A key moment when it surfaced was in 1901, by reference to the problem of how to determine the precise moment of contact during the transit of Venus. In a discussion where Bergson sought to emphasize the constructed nature of our knowledge of physical phenomena, the mathematician Louis Couturat raised the counterexample of the transit of Venus. "An eclipse, or even better, the transit of Venus across the sun," he argued, was proof that some physical phenomena were highly precise and delimited events. Bergson disagreed, insisting that physical phenomena were *never* naturally delimited: "It is the astronomer that," with his instruments, "*catches* the position of the planet from the continuous curve it traverses."[23]

Rudimentary precinematographic cameras first used during the transit of Venus were greatly improved in the years that followed. In light of these improvements Bergson's critique grew even bolder and sharper. In *Creative Evolution,* Bergson expanded this idea further. It was not only Venus's form—even under the scrutiny of Janssen's revolver—that was elusive, but all forms: "there is no form, since form is immobile and reality is movement. What is real is the continual *change of* form: *form is only a snapshot view of transition.*"[24]

---

22. Max Planck, *Eight Lectures on Theoretical Physics,* trans. A. P. Willis (Mineola, N.Y.: Dover, 1998).
23. "C'est l'astronome qui *cueille* cette position de l'astre sur la continuité de la courbe qu'il décrit." Henri Bergson, "Le Parallélisme psycho-physique et la métaphysique positive," in *Mélanges* (Paris: Presses Universitaires de France, 1972 [1901]), 502.
24. Ibid., , 502, and *Creative Evolution* (Mineola, N.Y.: Dover, 1998), 302.

Bergson found the "cinematographic method" both pervasive and constraining. Referring not merely to the modern cinematographic camera, but to the proclivity of the human mind for arranging temporal images spatially, he criticized its restrictiveness and urged scientists to "set the cinematographical method aside" and search instead for a "second kind of knowledge."[25] The philosopher campaigned against the facile, cinematographic distinction between the discrete and the continuous in life and in logic. He forged an influential philosophical movement with important repercussions for both science and philosophy. The renowned philosopher William James, for example, claimed that Bergson had compelled him to "give up the logic, fairly, squarely and irrevocably."[26]

By the first decade of the twentieth century Bergson had reversed the orthodox relation of form to movement: reality was primarily determined by movement. This perspective stood in contrast to the one described in the 1830s by the Belgian scientist Joseph Plateau who developed popular phenakistiscopic instruments that gave to static images the illusion of movement. Plateau argued that these instruments rescued form from the movement that obscured it: "One could in that way . . . *obtain the real form* of objects whose speed prevents us from distinguishing."[27] Yet for Bergson, the relation was simply the reverse: movement was real while forms were simply derivative.[28]

On numerous occasions Bergson explained how cinematography provided viewers with an illusory kind of movement that differed from real movement: "Suppose we wish to portray on a screen a living picture . . . how could it, at its best, reproduce the suppleness and variety of life?" By taking "a series of snapshots" and projecting "these instantaneous views on the screen, so that they replace each other very rapidly," movement could be reproduced. Yet this *illusory movement* resulted from shifting *real movement* elsewhere. In the case of cinema, it resulted from shifting it inside the apparatus: "In order that the pictures may be animated, there must be movement somewhere. The movement does indeed exist here; it is in the apparatus."[29] Bergson exhorted his followers to peer into cinematographic

25. Bergson, *Creative Evolution*, 342.
26. William James, *A Pluralistic Universe* (London: Macmillan, 1909), 226, cited in Paul Douglass, "Bergson and Cinema: Friends or Foes," in *The New Bergson*, ed. John Mullarkey (Manchester: Manchester University Press, 1999), 219.
27. Cited in Louis Olivier, "Physiologie: La Photographie du mouvement," *Revue scientifique* 30 (1882): 806. Italics mine.
28. For the historical precedents of the relation between form and movement, see Jimena Canales, "Movement before Cinematography: The High-Speed Qualities of Sentiment," *Journal of Visual Culture* 5, no. 3 (2006).
29. Bergson, *Creative Evolution*, 342.

black boxes to find real movement hidden there—to find, inside, a moving universe that could never be divided into separate, constitutive elements.

The central point was that no instrument—no matter how perfect—could escape the effects of time. He concluded his discussion with Einstein by stressing just this aspect of his philosophy. Although he "agreed entirely with M. Piéron," who had intervened in the debate with the example of the personal equation, that "the psychological determination of simultaneity is imprecise," he went further. Even laboratory determinations could not escape from this imprecision: "What is more, to establish this point by reference to laboratory experiments, it is also here that we must have recourse to psychological determinations of simultaneity—still imprecise. Without them it would be impossible to read an instrument."[30] Laboratory instruments, he reminded his listeners, had to be read. And these reading practices reintroduced psychological and philosophical riddles. While some physicists, such as Einstein, were willing to overlook the complex processes involved in reading instruments, for the philosopher they were essential.

Einstein fought against Bergson's interpretation. In the years that followed their notorious debate, he succeeded in dissociating his theory of relativity from reaction time research to the point that subsequent commentators have completely forgotten the intricacies of their relation. He was largely successful. After the encounter between them, it became increasingly difficult to think about physics in relation to reaction time and thus to examine the relation between physics and psychology. Most importantly, it became increasingly harder to think about science in relation to everyday practices involving movement and reaction. Time, clocks, and simultaneity were important for Bergson because of the role they played in everyday human affairs: "If this simultaneity did not exist, clocks would serve no purpose. They would not be manufactured, or at least no one would buy them. Because one buys them to know what time it is and 'to know what time it is' consists in finding a correspondence, not between one clock and another one, but between one clock and the present moment, the event which just passed, or anything which is external to the clock."[31]

Duration and Simultaneity

When Bergson's book appeared, the debate between the philosopher and the physicist only intensified. According to the theory of relativity, two

30. Bergson, "Discussion avec Einstein," 1344.
31. Ibid.

twins, one who traveled outside the earth at a speed close to that of the speed of light and the other who remained on the earth, would meet each other and notice that time had elapsed differently for each of them. Their clocks and calendars would show disagreeing times and dates. The twin who had stayed on the earth would have aged more rapidly; time would have slowed down for the one who had traveled.

In his controversial book Bergson denied that this was the case. He categorically stated that the clock of the traveling twin "does not present a retardation when it finds the real clock, upon its return."[32] Critics since then have often cited Bergson's remark that "once reentering, it marks the same time as the other" as proof of his profound misunderstandings of relativity.[33] This single statement was enough to discredit him in the eyes of most scientists—and it remains the case to this day.

But Bergson's controversial statement was part of a much larger argument that has been forgotten. In fact, Bergson *did* acknowledge that the twins' times would differ under most circumstances. His statement only held under quite special circumstances—circumstances that did not allow for *any* differences in the twins' situations, not even differences in acceleration. Explicitly focusing only on movement that was "straight and uniform," he demanded that "their situations be identical."[34] In every other case, Bergson accepted that the twin's clock times would differ.

In the first appendix to the second edition of his book Bergson expressed his irritation with readers who overlooked this aspect of his argument and who claimed that he denied the retardation of the traveling clock. He tried to prove them wrong by clearly stating his belief in relativity's effects on time: "We have already said it, and cannot cease to repeat it: in the theory of Relativity the slowing-down of clocks by their displacement is *completely evident* and *as real* as the shrinkage of objects in terms of distance."[35] But few listened.

Since then, Bergson has frequently been considered to have held scientific facts in disdain. Yet even in the preface to the first edition of *Duration and Simultaneity,* he showed every possible respect for the facts of observation: "we take the formulas of Lorentz, term by term, and we find out to which concrete reality, to what thing perceived or perceptible, each term

32. Appendix III in Henri Bergson, "Durée et simultanéité: À propos de la théorie d'Einstein," in *Mélanges* (Paris: Presses Universitaires de France, 1972 [1922]), 238.
33. Alan Sokal and Jean Bricmont, "Un Regard sur l'histoire des rapports entre science et philosophie: Bergson et ses successeurs," in *Impostures intellectuelles* (Paris: Odile Jacob, 1997), 176.
34. Bergson, "Durée et simultanéité: À propos de la théorie d'Einstein," 128.
35. Ibid., 216. Italics mine.

corresponds."[36] Bergson, who knew Hendrik Lorentz and Albert A. Michelson personally and deeply admired them, wanted *more* not *less* weight placed on Lorentz's formulas and on the results of the Michelson-Morley experiment.[37] He complained that "the Theory of Relativity is not precisely based on the Michelson and Morley experiment."[38] Einstein, in contrast, showed a surprisingly cavalier unconcern for its experimental results.[39]

Bergson applied to Einsteinian relativity an objection once brought up by the Cambridge philosopher Henry More in the seventeenth century against Descartes' theory of relative motion.[40] Using the example of a ship leaving its dock, Descartes argued that there was no way of distinguishing absolute from relative motion. Henry More objected, in the same way Bergson would later object to Einstein, that a certain difference would always remain between the two motions. More reminded Descartes that if one person was at rest (seated) and the other in motion (running), their differences could be clearly determined because the person doing the running would be flushed. For Bergson, acceleration was tantamount to being flushed from running, as in More's example. It was an inescapable mark of a difference in the twins' situations. Since *a* difference existed, one that resulted in a difference in times, then their times were not equal in *every* respect. Essential differences, for example, could remain in terms of memory or in a sense of effort.

Differences in clock times, arising in connection to differences in acceleration, proved that something in one of the twin's experience of time was different from the other's. Acceleration created a dissymmetry that in turn proved that the twins' times were not equal in every sense: "So, if one wants to deal with Real Times then acceleration should not create a dissymmetry, and if one wants for the acceleration of one of these two systems to effectively create a dissymmetry between them, then we are no longer dealing with Real Times."[41] A change in acceleration was necessary for the traveling twin to change directions and return to Earth to compare his time. Under these circumstances their experiences of time were thus not *entirely* identical.

36. In the preface to the first edition of ibid., 59.
37. For his meeting and impression of Michelson, see Bergson to Xavier Léon, Paris, 8 May 1921 in Henri Bergson, *Correspondances,* ed. André Robinet (Paris: Presses Universitaires de France, 2002), 934.
38. Bergson, "Durée et simultanéité: À propos de la théorie d'Einstein," 63.
39. The classic work advancing this thesis is Gerald Holton, "Einstein, Michelson, and the 'Crucial' Experiment," *Isis* 60 (1969): 133–297.
40. Descartes, *Principes,* vol. 2 (1644); and H. Morus, *Scripta Philosophica* (1679), vol. 2.
41. Henri Bergson, "Les Temps fictifs et les temps réel" (1924), in *Mélanges* (Paris: Presses Universitaires de France, 1972), 1444.

Bergson disagreed with Einstein about what would happen once the twins met back on Earth. The philosopher André Lalande, who wrote about the debate, explained: "The chief question here, of course, is to know what sort of *reality* should be accorded to the various opposed observers who disagree in their measurement of time."[42] Although *physically* the twins' times were equally valid, *philosophically* differences could remain between them. Whose time would prevail back on Earth would depend on how their disagreement was negotiated—not only scientifically, but psychologically, socially, politically, and philosophically. More's ironic thesis became for Bergson a seventeenth-century version of the twin paradox. It encapsulated Bergson's main point: that philosophy had the right to explore the differences in time and distance that relativity had shown varied among observers.

The Time of Their Lives

After meeting for the first time in Paris, Bergson and Einstein were scheduled to meet again in a few months, this time, for an entirely different purpose. Bergson was president of the International Commission for Intellectual Cooperation of the League of Nations, a forerunner of UNESCO, and Einstein was named as one of its members. Bergson was well aware that the power of the Commission depended on the strength of its members.[43] While the participation of both men already augmented the prestige of the Commission, the excitement around it only intensified after their heated encounter in Paris. The fate of the Commission was now colored by the Bergson-Einstein debate. For its participants, it was at least as important as the meeting itself.

Bergson's choice for the presidency of the Commission for Intellectual Cooperation was obvious, since he was the single most politically committed intellectual of his time. In 1916 he had gone to Spain on a diplomatic mission to try to have it side with the allies. The next year he played an essential role in convincing Wilson to enter the war against Germany. His reputation as a mender of intellectual schisms was established during his tenure as president of the Académie des sciences morales et politiques. At the start of the Great War, a group of members of the Institut de France demanded the expulsion of foreign associates of German nationality.[44] The

42. André Lalande, "Philosophy in France, 1922–1923," *The Philosophical Review* 33, no. 6 (1924): 543.
43. Jean-Jacques Renoliet, *L'UNESCO oubliée: La Société des nations et la coopération intellectuelle (1919–1946)* (Paris: Publications de la Sorbonne, 1999), 31.
44. For the banning of the Germans from various international unions after the war, see Daniel J. Kevles, "'Into Hostile Political Camps': The Reorganization of International Science in World War I," *Isis* 62

philosophers of the Institut, as a group, condemned this initiative. Bergson was responsible for drafting a declaration that condemned the war but did not go to the extreme of expulsing German nationals.[45]

Einstein was well aware of how politics affected science. He, for example, boycotted the Solvay Congresses that were resumed after the war in order to protest the exclusion of German scientists from international scientific forums.[46] He knew that his internationalist stance made him popular with some, unpopular with others. During those years he became deeply concerned with expanding the relevance of his theory of relativity *beyond* the community of physicists. In 1916 he published a "gemeinverständlich" version of both the special and general theory. His catapulting to fame by Eddington's eclipse expedition soon dwarfed these early popularization attempts. Popular expositions of relativity proliferated almost automatically after this date. His *Über die spezielle und die allgemeine Relativitätstheorie (gemeinverständlich)* was translated into English, French, Spanish, and Italian. Then came his famous Four Lectures on Relativity, presented at Princeton University in 1921.

His newfound fame was loaded with political meaning, representing the triumph of internationalist (allied) science over base nationalist (German) passions. Almost immediately after he joined the Commission for Intellectual Cooperation in 1922 Einstein thought of resigning because of the prevailing anti-German sentiments of many of its members.[47] He did not feel he could adequately represent Germany because of his "condition as a Jew, on the one hand, and on the other because of his anti-chauvinistic feelings from the German point of view do not permit him to truly represent the intellectual milieu and the Universities of Germany."[48] The physicist and chemist Marie Curie and others pleaded with him, and he chose to remain. But his support for the Commission remained so lukewarm that he missed its first meeting (1–5 August 1922).[49] Soon thereafter (21 March 1923), he resigned in earnest, publishing a sharply worded statement against it.

(1971). In 1926 the International Research Council (IRC) decided to stop the exclusion and invited German scientists to join as members. The Germans, however, declined.

45. This incident is described in Bergson, *Mélanges*, 1104. And in the introduction to Angelo Genovesi, "Henry Bergson: Lettere a Einstein," *Filosofia* 49, no. 1 (1998): 12 n. 2.

46. He decided to resume attending the Solvay Congresses in April 1926.

47. Germany did not initially belong to the League of Nations, which was created in Paris in the aftermath of the 18 April 1919 Peace Conference to regulate disputes among nations. Germany joined in 1926 after Locarno (1925). Einstein's claim that they were excluding all Germans provoked Lorentz to write a letter (on 15 September 1923) explaining to him that it was not true that German scientists were excluded on principle. On Einstein's vacillations about his membership see Nathan and Norden, eds., *Einstein on Peace*, 59.

48. Cited in Renoliet, *L'UNESCO oubliée: La Société des nations et la coopération intellectuelle (1919–1946)*, 27.

49. For Einstein's absence from the 13 September 1922 meeting, see Commission internationale de coopération intellectuelle, in *Mélanges* (Paris: Presses Universitaires de France, 1922), 1353.

Einstein accused the Commission of being "even worse" than the League of Nations and of appointing "members whom it knew to stand for tendencies the very reverse of those they were bound in duty to advance."[50] His highly publicized resignation only made the work of these institutions more difficult. His behavior appeared paradoxical to many of his colleagues.[51] How could a scientist who preached internationalism refuse to participate in these outreach activities? After all, he had been invited (they had indeed pleaded) as a German-born member. Had not Einstein repeatedly protest the exclusion of German scientists?

During this tumultuous period Einstein framed his theory of relativity and his debate with Bergson symmetrically, on both political and scientific grounds. In a letter to his friend Maurice Solovine, he connected his decision to resign from the Commission for Intellectual Cooperation to Bergson's reception of relativity: "I resigned from a commission of the League of Nations, for I no longer have any confidence in this institution. That provoked some animosity, but I am glad that I did it. One must shy away from deceptive undertakings, even when they bear a high-sounding name. Bergson, in his book on the theory of relativity, made some serious blunders; may God forgive him."[52] Forced on other occasions to explain his decision to resign while combating views that he was being pro-German, he again stated that his position with regard to the Commission for Intellectual Cooperation was consistent with the theory of relativity. In a letter to Marie Curie written in December of 1923, he explained: "Do not think for a moment that I consider my own fellow countrymen superior and that I misunderstand the others—that would scarcely be consistent with the Theory of Relativity."[53] Relativity, in those years, went far beyond his famous 1905 and 1918 *Annalen der Physik* papers. To Einstein and to those who followed his relationship with Bergson and the Commission, it encapsulated distinct

50. Einstein to Albert Dufour-Feronce (from the German Foreign Office), 1923, 3 pp., Document No. 34 877, in Folder 10, Folders 34.9–11 League of Nations, Box 14, Einstein Archives (hereafter cited as EA). Published in Albert Einstein, *Mein Weltbild* (Amsterdam: Querido, 1934). Reprinted in Albert Einstein, *Ideas and Opinions* (New York: Three Rivers, 1984), 84–85, on p. 85.

51. See, for example, the reaction of his colleague Max Born: "The papers report that you have turned your back on the League of Nations. I would like to know if this is true. It is, indeed, almost impossible to arrive at any rational opinion about political matters, as the truth is systematically being distorted during wartime." Born to Einstein, Göttingen, 7 April 1923, in *The Born-Einstein Letters: Friendship, Politics and Physics in Uncertain Times* (New York: Macmillan, 2005).

52. Einstein to Solovine, Pentecost 1923 (20 May 1923) in Albert Einstein, *Letters to Solovine* (New York: Philosophical Library, 1987), 58–59, on p. 59. Translated as "Bergson, dans son livre sur la théorie de la relativité, a fait des boulettes monstres; Dieu lui pardonnera," in Jean-Marc Lévy-Leblond, "Le Boulet d'Einstein et les boulettes de Bergson," ed. Frédéric Worms *Annales bergsonniennes* 3 (2007): 238. The last sentence is cited as "God forgive him" in Abraham Pais, *Einstein Lived Here* (Oxford: Clarendon Press, 1994), 75.

53. The letter appears in Carl Seelig, ed., *Albert Einstein: Eine dokumentarische Biographie* (Zürich: Europa Verlag, 1954), 209–10, on p. 210. It is also cited in Nathan and Norden, eds., *Einstein on Peace*, 64–65.

political and ethical views. With his growing international fame Einstein started to become more than a physicist. He adopted a role that would remain with him up to this day, that of a physicist-philosopher, with vocal political opinions.

Bergson temporarily had the last word during their meeting at the Société française de philosophie. His intervention negatively affected Einstein's Nobel Prize, which was given "for his services to theoretical physics, and especially for his discovery of the law of the photoelectric effect" and not for relativity. The reasons behind this decision were stated in the prize's presentation speech: "Most discussion [of Einstein's work] centers on his theory of relativity. This *pertains to epistemology* and has therefore been the subject of lively debate in philosophical circles. It will be no secret that the famous philosopher Bergson in Paris has challenged this theory, while other philosophers have acclaimed it wholeheartedly."[54] For a moment, their debate dragged the theory of relativity out of the solid terrain of science and into the shaky ground of epistemology.

But Einstein and his followers in Paris did not permit the philosopher's confrontation to pass so lightly. Within Paris, divisions between physicists and philosophers colored Einstein's reception. He was embraced by the Collège de France (particularly by Paul Langevin, who had invited him), greeted at the border by an astronomer from the Paris Observatory (Charles Nordmann met Einstein along with Langevin), courted by the Société de philosophie (in which forum he debated with Bergson), admired at the Société astronomique de France (especially by its president, the prince Bonaparte), and welcomed by the Société de chimie physique. The Société française de physique, ironically, rejected him completely.[55]

Jean Becquerel, the son of the eminent physicist Henri Becquerel, defended Einstein and attacked Bergson. Becquerel was the first physicist after Langevin to introduce classes on relativity at the École polytechnique and at the Muséum d'histoire naturelle, where he was a professor. He published two books on relativity in 1922, one of them designed for a general audience. In an article published in the *Bulletin scientifique des étudiants de Paris,* he took up the fight against *le bergsonisme.*[56] Becquerel insisted that

54. Arrhenius, 10 December 1922, in *Nobel Lectures in Physics, 1901–21* (New York: Elsevier, 1967). Cited in Pais, *Einstein Lived Here,* 75. Italics mine.
55. See note 20.
56. Jean Becquerel, "Critique par M. J. Becquerel de l'ouvrage Durée et simultanéité de M. Bergson," *Bulletin scientifique des étudiants de Paris* (March 1923).

objections against Einstein's theory resulted from misunderstandings and erroneous reasoning.[57]

This attack was followed by one in a book by André Metz, a military man and alumnus of the École polytechnique, who in 1923 published yet another book on relativity.[58] In all of these publications he attacked the philosopher, guilty for transforming a beautiful "child" into a "monster."[59] Bergson replied to these objections in a new preface and three appendices to the second edition of *Duration and Simultaneity*.[60] These appendixes, however, only intensified the debate between him and the physicists. In 1924 Metz wrote a direct response to Bergson's new works.[61]

Bergson responded once again to Metz in an article titled "Les Temps fictifs et les temps réel" (May 1924) in which he again tried to defend his philosophy.[62] He countered Metz's claim that he was professing a relativity theory that differed from Einstein's. All he was doing, he insisted, was philosophy—*not physics,* and these two disciplines were different: "Toute autre est le rôle du philosophe."[63] Metz's claim that physicists had a "special competence" with respect to questions of time and relativity was therefore inapplicable. And physicists, he added, were rarely philosophically competent: "one can be an eminent physicist and not be trained to the handling of philosophical ideas . . . it is in vain that one argues here their special competence: the question no longer belongs to physics." He chastised relativity theory for its desire to "stop being a physics to become a philosophy."[64] Bergson felt that questions of authority were being brought up gratuitously. Reacting against a perceived growth in the authority of physicists, he concluded: "Besides, whether we are dealing here with physics or philosophy,

57. Jean Becquerel, "Préface," in *La Relativité: Exposé sans formules des théories d'Einstein et réfutation des erreurs contenues dans les ouvrages les plus notoires* (Paris: Etienne Chiron, 1923), vi.

58. André Metz, *La Relativité: Exposé sans formules des théories d'Einstein et réfutation des erreurs contenues dans les ouvrages les plus notoires* (Paris: Etienne Chiron, 1923).

59. André Metz, *Temps, espace, relativité,* Science et Philosophie (Paris: Gabriel Beauchesne, 1928), 2:13.

60. Henri Gouhier, "Avant-Propos," in *Mélanges* (Paris: Presses Universitaires de France, 1972).

61. Metz's response first appeared in the *Revue de philosophie,* vol. 31 (1924). In a letter to Einstein he explained its purpose: "L'article de la Revue de philosophie à pour but de refuter Bergson." See Metz to Einstein, 30 April 1924, Bonn to Berlin, 4 pp., Document No. 18 250, Folder 18.8. A. Metz, Paris(2), Box 7, EA. Bergson's published exchanges with Metz were reprinted in Gunter, ed., *Bergson and the Evolution of Physics,* 135–90.

62. First published as Henri Bergson, "Les Temps fictifs et les temps réel," *Revue de philosophie* 24, no. 3 (1924).

63. Bergson, "Durée et simultanéité: À propos de la théorie d'Einstein," 232–33. "Physics could do a service to philosophy by abandoning certain ways of speaking that induce philosophy into error, and that risk confusing physicists themselves about the metaphysical significance of their views." Appendix III of Bergson, "Durée et simultanéité: À propos de la théorie d'Einstein," 237.

64. Bergson, "Durée et simultanéité: À propos de la théorie d'Einstein," 64.

the recourse to authority has no value."[65] Ultimately, he accused Metz of not having understood him.[66]

Metz was not alone in ignoring Bergson's insistence that he was doing philosophy—*not* physics. The strategy was in fact due to Einstein. In a private letter to Metz, he echoed the judgment made previously to Solovine with regards to Bergson's mistake. This time he wrote: "It is regrettable that Bergson should be so thoroughly mistaken, and his error *is really of a purely physical nature,* apart from any disagreement between philosophical schools." He spelled out Bergson's mistake in detail: "Bergson forgets that the simultaneity . . . of two events which affect one and the same being is something absolute, independent of the system chosen."[67] The director of the prestigious *Revue de philosophie* did not hesitate to publish Einstein's response.[68]

The letter Einstein sent to Metz arguing that Bergson's error was due to his misunderstanding of physics was not the only one sent and not the only one subsequently published. At least one other reached Miguel Masriera Rubio, a professor of physical chemistry in Barcelona. Masriera Rubio became Einstein's defender and Bergson's attacker in the Spanish-speaking world.[69] In articles published in the prestigious *Vanguardia* newspaper, he brought the debate to the public. Like Metz, Masriera Rubio decided to publish a letter from Einstein that contained the following damning sentence: "In short, Bergson forgets that spacetime simultaneity has an absolute character according to the Theory of Relativity."[70] With these two letters and their dissemination, Einstein effectively ended the controversy in his favor.

65. Bergson, "Les Temps fictifs et les temps réel," 1439.

66. "The meaning of my thoughts, as that of my book, has completely escaped him. There is nothing I can do." Henri Bergson, "Bergson à E. Peillaube," *Revue de philosophie* 24 (July 1924): 440. Reprinted in Bergson, *Mélanges,* 1450.

67. Einstein to Metz, 2 July 1924, republished in Gunter, ed., *Bergson and the Evolution of Physics,* 190.

68. La Direction, "Réponse de Henri Bergson," *Revue de philosophie* 24 (July 1924). Metz informed Einstein of its publication: "Il est bien entendu que la lettre que je sollicite de vous est destinée à être insérée (sauf avis contraire de votre part) dans la Revue de Philosophie.—J'ai d'ailleurs fait bien souvent usage de vos lettres (qui m'ont servi à dissiper bien de erreurs) et j'espère que vous n'y voyez pas d'inconvénient." See Metz to Einstein, 23 November 1924, Bonn, 5 pp. on p. 2, Document No. 18 253, Folder 18.8. A. Metz, Paris (5), Box 7, EA.

69. See Masriera Rubio, "La verdad sobre Einstein," 5 January 1925, Document No. 17 086; "El antirrelativismo psicológico," 7 January 1925, Document No. 17 086.1; "El estado actual de las doctrinas de Einstein," 25 October 1925, Document No. 17 088 in Folder 3 of 3, Miguel Masriera (1) Scientific Correspondence File, Folders M-Misc.1, Box 6, EA.

70. "Kurz: Bergson vergass, dass *raumzeitliche* Koinzidenz auch nach der Relativitätstheorie absoluten Character hat." Einstein to Masriera Rubio, 7 October 1925, Berlin, 2 pp. on p. 1, Document No. 17 087 in Folder 3 of 3, Miguel Masriera (2) Scientific Correspondence File, Folders M-Misc.1, Box 6, EA. The letter was republished in translation in Masriera Rubio, "De Einstein a mis lectores . . . ," *La Vanguardia,* November 1925, 3 pp., see Document No. 17 089, in Folder 3 of 3, Miguel Masriera (3) Scientific Correspondence File, Folders M-Misc.1, Box 6, EA.

Why, despite Bergson's repeated claims that he fully accepted the physics of relativity, and that he was only doing philosophy, did Einstein (through Metz, Masriera Rubio, and others) insist on Bergson's incompetence as a physicist? Bergson speculated that the reason was that Einstein simply did not understand him. In a letter to Lorentz, he explained: "In general, relativity physicists have misunderstood me. They, by the way, frequently do not know my views except through hearsay, by inexact and even completely false accounts. This is perhaps the case of Einstein himself, if what they say about him is true."[71] Framing the terms of the debate in terms of physics had two consequences: First, it denied Bergsonian philosophy the right to mark its independence from physics. Second, it assigned to physicists a "special competence" with respect to these questions.

In the end, Bergson and Einstein's debate could be summarized as a disagreement about how one should deal with disagreement. This included disagreement about time coordination, as in the example of the twins, but included other issues. Should one deal with disagreement as a physicist or as a philosopher? Through negotiation or by fiat? As an expert or as a commoner? These questions were all pertinent at the Commission for Intellectual Cooperation, where both Bergson and Einstein were trying hard to work for peace.

Discussions of time were particularly relevant for the Commission for Intellectual Cooperation for one essential reason. Its organization was modeled after previous scientific international commissions created for global sciences (such as geodesy and meteorology), global industries (such as electric, telegraphic, and rail), and global standards (time, longitudes, weights, and measures).[72] In these forums physicists, astronomers, and even engineers waxed philosophical, pondering, in universal terms, about the nature of science, consensus, and truth. The famous scientist Henri Poincaré, for example, developed his philosophy of conventionalism in the context of international debates about the standardization of time and longitude. His position contrasted with that of Einstein who never worked through the

---

71. Bergson to Lorentz, Paris, 9 November 1924 in Bergson, *Correspondances*, 1119–22, on p. 22.
72. One of the most important associations of this kind was the International Association of Academies (IAA) founded in 1899 and counting 22 members by 1914. After the treaty of Versailles the International Research Council (IRC) was set up. The International Council for Science (ICSU) was founded in 1931. For the history of the IAA, see Brigitte Schroeder-Gudehus, "Les congrès scientifiques et la politique de coopération internationale des académies de sciences," *Relations Internationales* 62 (1990). For the IRC, see Kevles, "'Into Hostile Political Camps': The Reorganization of International Science in World War I." For the ICSU, see H. Spencer Jones, "The Early History of the ICSU 1919–1946," *ICSU Review* 2 (1960).

long, painful negotiations necessary for reaching an agreement on time standardization.[73]

Perhaps Bergson was pessimistic. He, after all, had seen France, Germany, and Britain engaged in a bitter debate about which country's time and time keeping methods would prevail. Would not the twins, in the famous paradox, have to do the same in order to live together on Earth, peacefully?

The Intellectual Cooperation Experiment

The debate between Einstein, Metz, and Bergson appeared in the July 1924 issue of the *Revue de philosophie*. That summer was colored by an equally painful debate at the Commission for Intellectual Cooperation. Should Einstein be asked to rejoin, even after he had sent to the press insulting remarks about the League of Nations? The question of reintegrating Einstein into the Commission resulted in part from the pressure of the British, who sought to profit from the diplomatic isolation of France brought about by her occupation of the Ruhr area and from the concurrent devaluation of the franc. Gilbert Murray (Commission member, scholar of ancient Greek literature, and world peace advocate) was afraid that without Einstein "this Committee, like all the organizations of the League of Nations, is in danger of having the Latin element overrepresented."[74] In a letter marked "confidential" he pleaded to Einstein and offered him a carefully worded statement aimed at combating public criticisms: "There would be no inconsistency in this. You resigned as a protest after the invasion of the Ruhr [in March 1923] and the subsequent embitterment of feeling between France and Germany, and your return to the Committee would mark the beginning of that rapprochement to which we are all looking forward."[75] Einstein accepted the offer and the official explanation.[76]

For the meeting *proper* (25 July 1924) Bergson reintroduced Einstein with a flattering speech, but during the meeting *break* their differences once again became evident.[77] The philosopher Isaak Benrubi, who decided to at-

73. See Peter Galison, "Einstein's Clock: The Place of Time," *Critical Inquiry* 26 (Winter 2000), and *Einstein's Clocks, Poincaré's Maps: Empires of Time* (New York: W. W. Norton and Co., 2003).
74. Gilbert Murray to Einstein, 10 July 1922, Oxford, 2 pp. on pp. 1–2, Document No. 34 777, Folder 9, Folders 34.9–11 League of Nations, Box 14, EA.
75. Murray to Einstein, 16 May 1924, Oxford Document No. 34 807, Folder 9, Folders 34.9–11 League of Nations, Box 14, EA.
76. For his reinvitation, see Eric Drummond (Secretary General to the League of Nations) to Einstein, 21 June 1924, Geneva, 1 p. Document No. 34 805, Folder 9, Folders 34.9–11 League of Nations, Box 14, EA.
77. Bergson, "Commission Internationale de Coopération Intellectuelle: Ouverture de la session," 24 July 1924, Archives de l'Unesco. Reprinted in Bergson, *Mélanges*, 1454–55.

tend the Commission for Intellectual Cooperation's meeting in Geneva only after learning that both Einstein and Bergson would attend, approached Einstein to ask him what he thought of *Duration and Simultaneity*.[78] Einstein offered his typical response, that Bergson had not understood the *physics* of relativity—that he had made a mistake.[79] Asked if he would continue the fight against Bergson, Einstein responded: "No, I do not intend to do that, unless Bergson himself provokes a polemic. But that would not help anybody." Einstein was willing to let bygones be bygones: "Es wird Gras darüber wachsen, und dann wird man mit mehr Objectivität darüber urteilen."[80]

Einstein and Bergson did not learn to work together at the Commission for Intellectual Cooperation. Passions again flared when the French government offered the Commission the option of building an International Institute for Intellectual Cooperation in Paris. Einstein (and others) expressed his concern that the Commission was international only nominally, while in effect controlled by France. Bergson tried to calm his fears, promising that the institute would remain "rigorously and completely international" even though he could not turn down the French government's generous offer.[81] This made Einstein more and more suspicious of the Commission's covert nationalism. He did not attend the next meeting, which was held in Paris, instead of in Geneva.[82]

While Einstein was suspicious of the Commission for Intellectual Cooperation's "internationalism," others were suspicious of Einstein's, particularly in light of his increasingly prominent role as a Zionist. Was it not contradictory, they asked, that Einstein was fighting for the establishment of Israel, while he was preaching about internationalism? For years Einstein maintained that these two endeavors were not contradictory: "my Zionism does not preclude cosmopolitan conceptions."[83] Bergson's response to anti-Semitism and the growing horrors of Nazism would be very different from Einstein's. His national affiliation always remained firmly

---

78. Isaak Benrubi, *Souvenirs sur Henri Bergson* (Neuchâtel: Delachaux & Niestlé, 1942).

79. Ibid., 107–8. Also in the Introduction of Genovesi, "Henry Bergson: Lettere a Einstein," 8–9. And in Rose-Marie Mossé-Bastide, *Bergson éducateur* (Paris: Presses Universitaires de France, 1955), 126.

80. Cited in Benrubi, *Souvenirs sur Henri Bergson*, 108. Also cited in Mossé-Bastide, *Bergson éducateur*, 126.

81. See Bergson to Einstein, 15 July 1925, St. Cergue, Switzerland, Document No. 34 814, Folder 9, Folders 34.9–11 League of Nations, Box 14, EA. Published in Bergson, *Correspondances*, 1161–62.

82. Mossé-Bastide, *Bergson éducateur*, 145.

83. Albert Einstein, "How I Became a Zionist," in *The Collected Papers of Albert Einstein* (Princeton, N.J.: Princeton University Press, 2002). Originally published as Albert Einstein, "How I Became a Zionist," *Jüdische Rundschau* (1921).

French. Einstein invited Bergson to participate in the inauguration of the Jerusalem University. Bergson declined in February 1925 saying he was too busy and quickly changed the topic back to Einstein's participation at the Commission.[84]

In August 1925, Einstein once again criticized the Commission as two-faced, and Bergson resigned, citing an illness.[85] Bergson's resignation from the Commission marked the end of his political involvement. After this date he completely retired from public life. The influence of the French intelligentsia in world affairs decreased in direct proportion to the decline in the health and prestige of its main proponent, Bergson.

Bergson dedicated the last years of his life to writing *Les Deux Sources de la morale et de la religion* (1932), a book whose pessimistic tone regarding war, peace, and cooperation was influenced by his own experience at the Commission. The failure of the League of Nations was not due to its powerlessness and its lack of an enforcement mechanism, as many believed. Its troubles ran much deeper: "Even if the League of Nations would take an armed form sufficient in its appearance . . . it would collide with the profound war instinct which covers civilization."[86] Einstein did not increase his attendance after Bergson's departure. From 1926 to 1930 he attended only three meetings.[87] The Commission lost momentum and had its last session on July 1939.[88]

Bergson's and Einstein's hopes for the Commission ended along with their debate. Its unraveling was as immediate as solutions were evident. A few days before the last meeting Gilbert Murray wrote to Einstein in a desperate attempt to solve the institution's woes: "The best solution of all these difficulties is obvious! It is that you should remain with us, but perhaps that is too much to hope for."[89] During the years that followed, the debate's participants suffered the consequences of a worsening political situation. In 1933 Einstein moved to America, abandoned his pacifism, and would soon start to advocate atomic bomb research. Metz's home was searched by the Germans who stole his correspondence with Einstein and who took the rest of his belongings. He immigrated to London joining Charles de Gaulle's

---

84. Bergson to Einstein, 5 February 1925, Published in Bergson, *Correspondances*, 1147. Catalogued as Document No. 34.26.00, EA.
85. Renoliet, *L'UNESCO oubliée: La Société des nations et la coopération intellectuelle (1919–1946)*, 72.
86. Henri Bergson, *Two Sources of Morality and Religion*, trans. R. Ashley Audra and Cloudesley Brereton (New York: Henry Holt and Co., 1935), 310.
87. Renoliet, *L'UNESCO oubliée: La Société des nations et la coopération intellectuelle (1919–1946)*, 93.
88. Ibid., 7.
89. Murray to Einstein, 21 June 1932, 2 pp. on p. 2, Document No. 34 892, Folder 10, Folders 34.9–11 League of Nations, Box 14, EA.

Free France resistance movement.[90] With Franco's rise to power, Masriera Rubio went into exile.

Bergson's response to Nazism was very different. After the fall of France to Nazi Germany in June 1940, he did not ask the German or Vichy government for special treatment. Renouncing all privileges he decided to wait his turn in line in the street and register with other French Jews, in the inclement December weather. He died on 3 January 1941.

Cinematographic Morality

Bergson continued to resist the physicist's interpretation of relativity up to the last years of his life. He also continued to think about short intervals of time. In *The Two Sources of Morality and Religion* he explained the difference between the "dynamic morality" that he advocated and the "static morality" that he despised by reference to the cinematographic camera. Static morality was one that "has become ingrained in customs, ideas, and institutions; its obligatory character is to be traced to nature's demand for a life in common." Dynamic morality, on the contrary, was "impetus, and . . . related to life in general, creative of nature which created the social demand." Just as real movement could not be obtained with static images that succeeded each other rapidly, so dynamic morality could not be reached through the fulfillment of static obligations. What was necessary was to break from "illusions we have time and again denounced": "the first, a very general one, consists in the conception of movement as a gradual diminution of the space between the position of the moving object, which is immobility, and its terminal point considered as reached, which is immobility also, whereas positions are but mental snapshots of the indivisible movement: whence the impossibility of reestablishing the true mobility, that is to say, in this case, the aspirations and pressures directly or indirectly constituting obligation."[91] In his very last work *La Pensée et le mouvant* (1934), the philosopher once again confronted the topic of cinematography. One of the reasons why the cinematographic camera did not reveal real movement was because "the film could pass ten times, one hundred times, one thou-

---

90. "In 1940–44 all my things were kept by the Nazis and disappeared from my flat in Strasbourg. I succeeded to go to England (with my sons) to join the Free French Forces. . . . I never found again the letters received from Einstein (and from Eddington, Becquerel, Meyerson, etc . . .)." Metz to Helene Dukas, 10 January 1960, Antony (Seine) to Princeton, Document No. 18 269, Folder 18.8. A. Metz, Box 7, EA. For her response and his interest in recovering Einstein's letters with "his opinion about Bergson," see their correspondence from Antony (Seine) to Princeton, Documents No. 18 270 and 18 271, Folder 18.8. A. Metz, Box 7, EA.
91. Bergson, *Two Sources of Morality and Religion*, 260.

sand times faster" and "that which passes would not be modified." What is more: "succession so defined does not add anything, it, in effect, removes something; it marks a deficit, it translates an infirmity of our perception."[92] To comprehend the nature of time, then, required understanding the short periods of time that produced the illusion of movement. These moments were important for Bergson because with them he could reattach questions of time to the human body, from the human body to its embeddings in the world, and from the world to ethics.

By thinking about short moments of time, Bergson advanced his famous critique of the cinematographic method. It was based on the conviction that something essential escaped from the small gaps, or frame lines, bordering successive film stills. In *Creative Evolution* (1907), he noted how nobody was looking at frame lines: "As to what happens in the interval between the moments, science is no more concerned with that than are our common intelligence, our senses and our language: it does not bear on the interval, but only on the extremities."[93] This "snapshot" mentality had profoundly negative ethical consequences. Bergson lamented that both common and scientific knowledge were completely restricted by it.

Echoes of Bergson's criticisms of the cinematographic method persisted in modern art and literature. Rebel turn-of-the-century artists, ranging from Rodin to the Italian futurists, stubbornly refused to employ cinematographic and even precinematographic technologies.[94] But in contrast to earlier critiques of cinematographic and precinematographic methods, the works of these modern artists were no longer legitimate contenders in the scientific

---

92. Bergson, "La Pensée et le mouvant," 1259–60.

93. Bergson, *Creative Evolution*, 330.

94. Rodin, intentionally, did not to conform his famous statue of St. John the Baptist to the rules of locomotion shown by chronophotography. He stubbornly claimed that "it is the artist who is veridical and the photograph which lies, since in reality time is not arrested." In his famous conversations with Paul Gsell, Rodin is credited with asking his friend, "Have you attentively examined the instantaneous photographs of walking men?" Rodin then proceeded to criticize them, saying how they looked "bizarre" and "paralyzed." He defended the work of Géricault who "they critique . . . because in his Course d'Epsom, which is at the Louvre, he has painted horses which gallop ventre à terre." See Auguste Rodin, *L'Art: Entretiens réunis par Paul Gsell* (Paris: Bernard Grasset, 1924), esp. 83–91. Seurat, who studied in the École at the same time Duval was teaching, continued painting the horse in the classic style as late as 1891; Baudelaire, famously, sided with Delacroix and his unphotographic portrayal of horses; Duchamp rebelled against snapshot instantaneity and showed movement through the use of sketchlike consecutive figures; the futurists also explicitly denounced traditional ways of portraying movement, particularly of their beloved speeding things. For the reaction of Rodin, Degas, and Seurat to chronophotography, see Marta Braun, *Picturing Time: The Work of Etienne-Jules Marey (1830–1904)* (Chicago: University of Chicago Press, 1992), 254, 272. For a commentary on the absence of chronophotographic gaits in the paintings of Seurat, see Jonathan Crary, *Suspensions of Perception: Attention, Spectacle, and Modern Culture* (Cambridge, Mass.: MIT Press, 1999), 276–78. For the futurists see Alexander Sturgis, *Telling Time* (London: National Gallery Co., 2000), 57.

debate surrounding the portrayal of movement.[95] These artists contrasted starkly with earlier researchers whose criticisms of the personal equation and their technological solutions formed part of standard scientific practices and established theories of knowledge. These criticisms were often formulated in the context of these pressing scientific problems, although posterity has almost exclusively remembered them as philosophical.

When the poet Paul Valéry received his copy of Bergson's last book, *La Pensée et le mouvant* (1934), he was particularly intrigued by a long footnote "on the subject of the *grande affaire* of Relativity." Referring to the recent advances in quantum mechanics that Einstein famously resisted, he asked if these "up-to-date microphysics" could be brought to bear on "some of your conceptions?"[96] Perhaps only a poet could hope for either reconciliation or for the comeback of *le bergsonisme*.[97]

What started out in 1922 as a minor disagreement between Bergson and Einstein about short moments of time—of the order of a tenth of a second—ended in major disagreements about much more: the role of physics versus philosophy, the twin paradox, the League of Nations. According to Bergson, this minor disagreement was directly connected to the other major ones. His philosophy connected them all together, and he urged his followers to change their views about short moments of time *first* in order to then change their views about the world. While it was commonly believed that certain instruments, ranging from automatic inscription devises to cinematographic ones, had completely demystified short moments of time by eliminating the errors that surfaced in their determination, Bergson offered a new way of thinking about these instruments and their relation to time. Yes, they solved some problems, but others inevitably escaped.

---

95. Rodin's interlocutor and young friend Paul Gsell could not help but accuse him, after hearing his reasons for rejecting chronophotographic poses, of not "copying Nature with the greatest sincerity." If anything, by virtue of their rebelliousness and their taste for shock effects the futurists and the avant-garde only reified the chronophotographic "rules" for portraying motion.

96. Paul Valéry to Bergson, Nice, 25 June 1934 in Bergson, *Mélanges*, 1511–12

97. In 1929 Einstein went to see Bergson with Paul Valéry. See Lévy-Leblond, "Le Boulet d'Einstein et les boulettes de Bergson."

CHAPTER 8

CONCLUSION

As to what happens *in the interval between the moments*, science is no more
concerned with that than are our common intelligence, our senses and our
language: it does not bear on the interval, but only on the extremities.

HENRI BERGSON, philosopher

It is an often-repeated truism that modern times are characterized by contradictions, particularly by a contradiction in our thinking about time. "The
new value placed on the transitory, the elusive and the ephemeral, the very
celebration of dynamism," explained Jürgen Habermas, "discloses a longing
for an undefiled, immaculate and stable present."[1] Michel Foucault saw a
similar tension in the era, which he traced to the poet Charles Baudelaire.
While Baudelaire described modernity as "the ephemeral, the fleeting, the
contingent," the philosopher reminded us how the period was also defined
by a desire of "recapturing something eternal."[2]

1. Jürgen Habermas, "Modernity versus Postmodernity," in *The Continental Aesthetics Reader*, ed. Clive
Cazeaux (London: Routledge, 2000), 269.
2. Charles Baudelaire, *The Painter of Modern Life*, trans. Jonathan Mayne (London: Phaidon, 1964), 13.
Cited in Michel Foucault, "What Is Enlightenment?" in *The Foucault Reader*, ed. Paul Rabinow (New York:
Pantheon, 1984).

A search for constants, absolutes, and universals came along with a fascination with the ephemeral, the fleeting, the unstable. The tenth of a second embodied this contradiction. In centuries marked by a constant search for faster and shorter moments of time, it remained a unit of salient importance. Appearing at the crossroads of a number of discourses in science, literature, and philosophy, it was both representative of the increasingly fast pace of modernity and a placeholder for an instant of stability. Modern culture held on to this moment and stressed its significance, even when the race to capture much shorter slices of time continued unabated.

Despite its often contradictory nature, the history of modernity is one of progress. The grand project of the Enlightenment proposed that science, understood in terms of empiricism and rationalism, was instrumental in solving some of the major problems of previous eras. Science gave this period its distinctive singularity, marking it off from darker ages and separating it from remote places still dominated by savagery and primitivism. As the tenth of a second became increasingly disenchanted during this period, researchers saw in this process the victory of "modern" scientific knowledge over other ways of knowing.

The success of modern science was guaranteed by rapidly advancing measurement techniques. During this period, measurement increasingly appeared to have a tight relation to social agreement and even justice. Through measurement we establish all sorts of equations—mercantile, scientific, and moral (for example, an eye for an eye, a tooth for a tooth). "Justice," Bergson reminded us, "is represented as holding the scales. Equity signifies equality. Rules and regulation, right and righteousness, are words which suggest a straight line. These references to arithmetic and geometry are characteristic of justice throughout its history."[3] The word "*Pensare,* from which we derive 'compensation' and 'recompense,' means to weigh."[4] Human justice emulated the justice of measurement. Measurement standards assured judgment standards. The physicist James Clerk Maxwell made the point plainly: "The man of business requires these standards for the sake of justice, the man of science requires them for the sake of truth, and it is the business of the state to see that our measures are maintained uniform."[5] Take the (1883) quotation by William Thomson, the well-known

---

3. "Justice has always evoked ideas of equality, of proportion, of compensation." Henri Bergson, *Two Sources of Morality and Religion*, trans. R. Ashley Audra and Cloudesley Brereton (New York: Henry Holt and Co., 1935), 60.
4. Ibid.
5. Maxwell, "Dimensions of Physical Quantities," Cambridge University Library MSS ADD 7655/Vh/4. Cited in Simon Schaffer, "Accurate Measurement Is an English Science," in *Values of Precision*, ed. M. Norton Wise (Princeton, N.J.: Princeton University Press, 1995).

physicist who would don the title of Lord Kelvin, who equated measurement with thought in a famous phrase, "when you can measure what you are speaking about, and express it in numbers, you know something about it; but when you cannot express it in numbers, your knowledge is of a meager and unsatisfactory kind."[6] In more recent work in the history of science, quantification still appears as the key scientific change during the Enlightenment.[7]

One of the main motivations behind tenth-of-a-second studies arose from their relation to measurement. Measurement practices and measurement-based science were often listed among the determining causes behind the great divide marking off modernity from other eras. They were even considered essential in the great divide between humans and nonhumans. Consider, for example, Nicolas of Cusa's *Idiota* (1450). In this text, quantification already appears as the main reason why men "excel beasts," for "beasts can neither number, weigh, nor measure." Or let me recount Voltaire's famous story "Micromégas," which, although published centuries later, expressed a similar idea. When a giant by the name of Micromégas from one of the "planets" surrounding Sirius and his much smaller, but still gigantic, friend from Saturn land on Earth and catch sight of the tiny beings populating the planet, they mistake them for animals. As the story develops, the ability to quantify appears as the determining factor behind the great divide between beings endowed with a "soul" and those without it.[8]

Measuring, mastering, and disenchanting the tenth of a second became an important challenge to the progress of modern science. A new understanding of stimulations and reactions as signals transmitted from instruments to the body, through the body, and from there back to the instruments was defined by reference to this value. This number was thus essential in

---

6. William Thomson, "Electrical Units of Measurement," in *Popular Lectures and Addresses* (London: Macmillan, 1891–94 [1883]), 73.

7. A "paradigmatic" example is Kuhn, "Mathematical versus Experimental Traditions in the Development of Physical Science," *Annales* (1975). Republished in Thomas S. Kuhn, "Mathematical versus Experimental Traditions in the Development of Physical Science," in *The Essential Tension: Selected Studies in Scientific Tradition and Change* (Chicago: University of Chicago Press, 1977).

8. Among these animals, they encounter some beings with whom they can communicate, and they proceed to ask them questions. To a question about the distance from the earth to the moon, the earthlings rightly respond: "Sixty times the radius of the Earth, in round numbers." Simply "amazed at their answers," Micromégas and the Saturnian acknowledge that these tiny earthlings are human because of their uncanny ability to measure. They were now "inclined to take as sorcerers the same people to whom a quarter of an hour earlier he had refused a soul." Accepting that they "are among the few who are enlightened," they learn that their interlocutors are "a flock of philosophers . . . just returning from the Arctic Circle," alluding to Maupertuis's expedition to measure the shape of the earth. Voltaire, "Micromégas: A Philosophical Story," in *Micromégas and Other Short Stories* (London: Penguin Books, 2002), 33.

negotiating the connection between "hardware" and "wetware" based on new concepts of communication and reference. It was decisive in establishing the clear-cut differences between body and spirit, humans and nonhumans, as well as nature and society—differences that helped frame a number of discourses in the post-Enlightenment era. For numerous scientists and thinkers, the elimination of tenth-of-a-second variations in measurement represented the ultimate achievement of modern science.

For almost a century, scientists invented new instruments, techniques, and practices in desperate attempts to control tenth-of-a-second moments and to eliminate errors due to slight variations in this value. With technology as a handmaiden, they solved most of these problems. Interferometers measured minute differences in length; impersonal micrometers followed stars in their passage across the sky and automatically recorded their movement; cinematography and high-speed photography were used to capture these fleeting moments. The nineteenth century's concern with tenth-of-a-second moments lost much of its vibrancy with the appearance of new instruments. Thus at the beginning of the twentieth century, scientists could celebrate that the previous century's obsession with the "sacred 0.1 second" had finally passed.[9]

Accounts of scientific practices have systematically ignored just what happens between these tenth-of-a second moments. While this lacuna could lead us to believe that they were unimportant, even a cursory look at basic scientific practices reveals the contrary. Gaining mastery over moments of this order, eliminating reaction time variations, and solving the problem of the personal equation were central problems confronting science and philosophy for nearly a century.

Short Durée History

Yet underneath this success story, problems of the order of the tenth of a second continued to rumble. Starting in the late 1950s historians started to question prevailing accounts of measurement in modernity. In most accounts measurement was seen as spreading from the exact sciences (primarily in astronomy) to conquer, by the beginning of the twentieth century, even psychology. The history of the personal equation confirmed this narrative, yet some thinkers seemed unconvinced and motivated to reconsider the personal equation and its standard history. Something in this tenth-of-a-second delay and something in the history of this problem still needed to

9. Edwin G. Boring, *A History of Experimental Psychology* (New York: Appleton-Century-Crofts, 1929), 148.

be explained. The case was reopened, becoming part of a broader inquiry into the nature of quantification.

One important milestone took place in 1959, during a Conference on the History of Quantification in the Sciences organized by Harry Woolf, editor of the history of science journal *Isis* and director of the Princeton Institute of Advanced Study. At the conference, Thomas Kuhn, a historian who at the time was preparing his famous *Structure of Scientific Revolutions* (first published in 1962), took issue against an earlier presentation at the conference: Boring's account of the history of the personal equation. He accused the psychologist of obscuring "what are perhaps the most important points" to be made about measurement. Boring, Kuhn protested, had an excessively broad understanding of measurement as any "scientific experiment or observation."[10]

In contrast to Boring's approach, Kuhn's limited his understanding of measurement to something that "always produces actual numbers." By zeroing in on measurement in a way ignored by his predecessors, Kuhn came to some surprising conclusions. Actual measurements rarely coincided with those predicted by theory. Kuhn mentioned an anecdotal "Fifth Law of Thermodynamics," often invoked by students doing laboratory work: "No experiment gives quite the expected numerical result."[11] Viewed in this way, progressive improvements in measurement techniques were not representative of scientific advances. Instead, Kuhn believed that bold theoretical presuppositions in the form of a "paradigm"—a word he famously coined—were more influential in shaping the development of science.

At the time Kuhn wrote his critique, most historians of science held a "positivist" view of science in which its grand theoretical structure rested on the collection of specific, uncontested observations. But because of the widely known yet frequently unacknowledged discrepancies that arose when scientists tried to measure something, Kuhn reversed the orthodox conception of science where theory rested on observation. He instead stressed the primacy of theoretical presuppositions: "the scientist often seems rather to be struggling with facts, trying to force them into conformity with a theory that he does not doubt."[12]

10. Harry Woolf, ed., *Quantification: A History of the Meaning of Measurement in the Natural and Social Sciences* (Indianapolis: Bobbs-Merrill, 1961). The essays of Boring and Kuhn are Thomas S. Kuhn, "The Function of Measurement in Modern Physical Science," in *Quantification: A History of the Meaning of Measurement in the Natural and Social Sciences*, ed. Harry Woolf (Indianapolis: Bobbs-Merrill, 1961); Edwin G. Boring, "The Beginning and Growth of Measurement in Psychology," in *Quantification: A History of the Meaning of Measurement in the Natural and Social Sciences*, ed. Harry Woolf (Indianapolis: Bobbs-Merrill, 1961).
11. Kuhn, "The Function of Measurement in Modern Physical Science," 35.
12. Ibid., 41.

The debate enveloping Boring's essay was part of a larger inquiry into the role of measurement in the progressive development of science. This theme was again an important discussion topic a few years later at the Symposium on the History of Science (1961). The historian of science I. Bernard Cohen chastised his colleague Maurice Daumas (from the Conservatoire national des arts et métiers) for claiming that the progress of science depended on improved measurement techniques brought about through the gradual improvement of precision instruments.[13] Daumas's thesis was widespread: most histories of science written during this period adopted this view of quantification. The standard account of the personal equation was based on this maxim: tenth-of-a-second errors having to do with individual differences or with reaction time were solved through instrumental improvements. They no longer posed a problem for scientists since new instruments allowed scientists to "translate the observation automatically into the visual reading of a scale."[14] Tamed by scales, these errors were significantly minimized.

The view that tenth-of-a-second variations were no longer meaningful in science was grounded on actual scientific practices. Exact scientists working during this period employed interferometers to eliminate the personal equation in measurements of lengths, astronomers increasingly employed impersonal micrometers, and numerous other scientists used direct-reading scales and automatic inscription devises.[15] Yet despite technological achievements that were implemented in observatories and laboratories, a few prominent historians disagreed with the view that considered precision solely in terms of improved instrumentation. It was "dangerous," according to I. Bernard Cohen, "to assume a simple one-to-one correspondence between increase in precision of measurement and the progress of science."[16]

13. See the continuing critique of Daumas's scholarship by Kathryn Olesko: "Recently historians have argued, as did Maurice Daumas over thirty years ago in an oft-cited and influential article, that in the late eighteenth and early nineteenth centuries more accurate measurements were the product of instrumentation." Kathryn M. Olesko, "The Meaning of Precision: The Exact Sensibility in Early Nineteenth-Century Germany," in *The Values of Precision*, ed. M. Norton Wise (Princeton, N.J.: Princeton University Press, 1995), 105.

14. Boring, "The Beginning and Growth of Measurement in Psychology," 118. "Such readings are the most acute and the least variable of all sensory discriminations."

15. During their study of the Salk Institute, Bruno Latour and Steve Woolgar estimated that scientists "spend two-thirds of their time working with large inscription devices." Bruno Latour and Steve Woolgar, *Laboratory Life: The Construction of Scientific Facts* (Princeton, N.J.: Princeton University Press, 1986), 69.

16. Maurice Daumas, "Precision of Measurement and Physical and Chemical Research in the Eighteenth Century," in *Scientific Change: Historical Studies in the Intellectual, Social and Technical Conditions for Scientific Discovery and Technical Invention, from Antiquity to the Present*, ed. A. C. Crombie (London: Heinemann, 1963); I. Bernard Cohen, "Commentary," in *Scientific Change: Historical Studies in the Intellectual, Social and Technical Conditions for Scientific Discovery and Technical Invention, from Antiquity to the Present*, ed. A. C. Crombie (London: Heinemann, 1963).

The lessons stressed by Kuhn and later historians reverberated decades later in another notable essay reevaluating the history of the personal equation: Simon Schaffer's often-cited article "Astronomers Mark Time." Like Kuhn before him, Schaffer took issue with Boring's essay. He criticized Boring's teleological assertions that the personal equation "was destined to become the property of the new physiological psychology" and that "at bottom the problem is psychological," taking the historian of psychology to task for introducing an argument that ignored how scientists themselves, in this case astronomers, dealt with the problem.[17] He profoundly disagreed with Boring's assertion that "experimental psychology can be said to have grown in part out of astronomy" and its implication that psychologists participated in the expansion of quantification by importing the methods of the physical sciences into the human sciences.[18] He sided instead with Kuhn, whose essay had appeared in the same volume as Boring's, standing by the claim that "it is impossible to separate processes of quantification from the preferred work styles which sustain them."[19] Schaffer's work belonged to a new brand of historiography that no longer considered quantification as linear, value free, universal, or good: "Quantification is not a self-evident or inevitable process in science's history, but possesses a remarkable cultural history of its own."[20]

After initially following Kuhn in stressing the importance played by theoretical concerns in guiding scientific research, historians and philosophers (including Schaffer), many of them associated with the sociology of scientific knowledge movement of the 1970s and 1980s, corrected what at times seemed an extreme anti-positivist tendency by returning to the study of scientific practices and experiments. A focus on sited, situated, instrumental, and experimental practices marked a "post-positivist" and "post-constructivist" philosophy and history of science that focused closely on local scientific cultures. These recent historians have reminded us that measurement, precision, and accuracy are never "purely technical." They have a rich social and cultural history of their own.[21] For example, different social groups compete to establish the metrological standards that suit their professional and class interests.

17. Boring, *A History of Experimental Psychology*, 133, 46.
18. Ibid., 142.
19. Simon Schaffer, "Astronomers Mark Time: Discipline and the Personal Equation," *Science in Context* 2 (1988): 118.
20. Ibid., 115.
21. M. Norton Wise, ed., *The Values of Precision* (Princeton, N.J.: Princeton University Press, 1995).

Implications for the Philosophy of Science

Sturdy philosophical justifications lay behind some of these new histories of quantification. If we look deep into them, we again see references to the tenth of second. If observations made by different observers were different, and if these differences were neither random nor easily averaged, how did our view of science need to be changed? Two philosophers, Karl Popper and Michael Polanyi, provided different answers to these questions. "Anti-positivist" work in the history of science was frequently supported by reference to their work.[22]

During his lifetime, Popper was best known for his theory of "falsificationism," proposing a view where science advanced when scientists crafted hypotheses that could be tested and potentially falsified by experiments. His outlook profoundly influenced the philosophy of science from the late 1950s onward, when science was frequently described in terms of theory formation followed by experimental corroboration. For recent historians, Popper's concept of falsificationism has been less important than his claims about the social nature of science—claims that underlined essential flaws in logical positivism. In contrast to most logical positivists, Popper placed a strong emphasis on the "social or public character of scientific method."

In addition to turning to Popper, "anti-positivist" historians of science have been inspired by the work of the chemist and philosopher Michael Polanyi, who stressed the role of trust and consensus in the formation of science. While both Popper and Polanyi have been widely cited, historians have forgotten that both of these philosophers turned to the personal equation as an important example for thinking about the development of science.

In his famous book *The Logic of Scientific Discovery* (1934), Popper used the example of the personal equation to illustrate how scientific hypotheses could never be unambiguously proved (at best, they could be falsified). While an observer could confidently make a statement about his direct experience of the world, Popper found no reason why statements of direct experience could not be infinitely questioned. This was evident with regard to the personal equation: "Thus it may become necessary, for example, to test the reaction-times of the experts who carry out the tests (*i.e.* to determine their personal equations)."[23] Since, in practice, most scientists stopped

22. See, for example, the use of Popper and Polanyi by Steven Shapin, *A Social History of Truth: Civility and Science in Seventeenth-Century England*, ed. David L. Hull, Science and Its Conceptual Foundations (Chicago: University of Chicago Press, 1994), 24.
23. Karl R. Popper, *The Logic of Scientific Discovery* (London: Routledge, 1959), 105.

testing their observations *after* investigating their personal equations, the personal equation was a practical, albeit dogmatic, place to stop. Popper admitted that "the basic statements at which we stop, which we decide to accepts as satisfactory, and as sufficiently tested, have admittedly the character of *dogmas*, but only in so far as we may desist from justifying them by further arguments (or by further tests)." He tempered his description of science as dogmatic by qualifying these dogmas as "innocuous" preconditions for building the "soaring edifice of Science," but his conclusion was nonetheless radical.[24] Science relied—in the final analysis—on dogmas.

Popper turned to the personal equation to make a "particularly obvious" point about science that has since fascinated historians: that its purported "objectivity" was really a type of "intersubjectivity." The personal equation was precisely the discovery illustrating the need to consider science as a shared practice. In the place of an "individual scientist's impartiality or objectivity" he claimed that "scientific objectivity can be described as the intersubjectivity of the scientific method."[25] Could Robinson Crusoe, castaway on an island, ever obtain scientific knowledge? What would happen, for example, if he discovered his personal equation? Could "Crusoe's discovery of his *'personal equation'* (for we must assume that he made this discovery), of the characteristic personal reaction-time affecting his astronomical observations" be considered a scientific discovery? No, according to Popper:

> In one point . . . is the . . . character of the Crusonian science particularly obvious; I mean Crusoe's discovery of his "personal equation" (for we must assume that he made this discovery), of the characteristic personal reaction-time affecting his astronomical observations. Of course, it is conceivable that he discovered, say, changes in his reaction-time, and that he was led, in this way, to make allowances for it. But if we compare this way of finding out about reaction-time, with the way in which it was discovered in "public" science—through the contradiction between the results of various observers—then the "revealed" character of Robinson Crusoe's science becomes manifest.

Thus, Crusoe's discovery could not be considered legitimately scientific until it became disseminated and accepted by the larger scientific community.[26]

The personal equation appeared as an example illustrating the social nature of science—as embodying the moment when the scientific community

24. Ibid., 104–5.
25. Karl R. Popper, *The Open Society and Its Enemies* (Princeton, N.J.: Princeton University Press, 1950), 403.
26. Ibid., 405. Italics mine.

agreed about what should count as "direct experience." For Popper it proved "that what we call 'scientific objectivity' is not a product of the individual scientist's impartiality, but a product of the social or public character of scientific method." He derived two lessons from the personal equation. First, it revealed that the moment at which scientists stopped questioning their observations was neither absolutely final nor permanently fixed. Second, it showed the inherently intersubjective character of science.

To Polanyi the personal equation brought an additional lesson. It proved the inescapable presence of personal judgment in science, that is, the "essential personal participation of the scientists even in the most exact operations of science."

Polanyi's research prompted the computer pioneer Alan M. Turing to recant some of his claims about the ability of computers to rival human intelligence. The debate between Polanyi and Turing turned, once again, to examples involving short moments of time. To build his case against artificial intelligence, Polanyi knew he "could substantiate [his argument] by quoting the Great Princeton astronomer H. N. Russell on the 'extremely troublesome errors' varying from observer to observer." But he chose instead "a more homely illustration." He discussed with Turing a photograph showing the final instant of a horse race. In the picture, one of the horses' heads appeared "a fraction of an inch ahead of another's." But the verdict as to the winner could not be established because "of the projection of a thick thread of saliva" that extended "forward by six inches or so well ahead of that of its rival." Polanyi explained how the decision on the winner had to be referred to the stewards of the Jockey Club. This example, according to him, made evident personal and interpretational aspects of science. This dimension of science could never be removed, not even with instrumental improvements or automatic recording mechanisms. While "the advent of the photo-finish camera . . . seemed to render the decision altogether obvious," in practice, this was not the case. Knowledge, he insisted, would always remain personal.[27] Instrumental improvements did not eliminate unforeseen elements that could arise at any moment, such as the one described in his suggestive example, which would always for call for personal interventions and judgments.

The lessons that Popper and Polanyi derived from the personal equation helped a generation of historians and philosophers rethink the process of knowledge production. This book continues that work, providing as well yet another account of the controversial events described by Ribot in the

27. Michael Polanyi, *Personal Knowledge* (Chicago: University of Chicago Press, 1958), 20 n. 1.

late nineteenth century, by Boring in the early twentieth century, and more recently by Schaffer and others.

Where Do We Go from Here?

Technical improvements did not solve tenth-of-a-second riddles on their own. Many other elements played essential roles. Some of these were institutional, such as the creation of laboratories of experimental psychology and the emergence of centers of calculation dedicated to metrology. These institutional changes were based on new divisions between disciplines (such as between the human and physical sciences) and within them, between physics and astronomy, physiology and psychology. They were marked by a shift in the status of astronomy as compared to physics when the latter grabbed the principal role in defining foundational metrical units. Scientific disciplines (like physics and physiology) that were once entangled became increasingly separated.

Other elements depended on new relations between science and the larger cultural sphere (illustrated by the different use of technologies in laboratories and in entertainment venues). The attempt to resolve human error was connected to the origins of techniques (like cinematography) that escaped from the laboratories where they were being deployed (not without dissent) as a neutral way to resolve discrepancies in scientific observation.

The acknowledgment of these errors affected how science was done. For example, debates about standardization and the benefits of certain publication practices arose as part of a larger debate about tenth-of-a-second errors. Additional changes were more revolutionary, such as new relations between science, art, and philosophy.

Some of these changes involved rethinking basic conceptions of the self. The tenth of a second was invoked in attempts to understand the difference between subjectivity and objectivity. Subjective perceptions, according to some authors, were marked by tenth-of-a-second limitations of the body, whereas objective ones overcame those same limitations. The importance of this value, however, was not limited to questions about universal human subjectivity, but it was used to investigate what was distinctly personal and individual about particular observations and about the testimony of these observations. What scientists learned after paying attention to this short moment of time was directly contrasted to the lessons previously learned from Descartes, Kant, and Hume.

We can also start to rethink aspects of knowledge production where the simple recipe of "experimental science" no longer suffices. Since antiquity, scientists argued for an active engagement with the world involving the aid of tools and instruments as a way to counter perceptual illusions. Even astronomers rarely relied exclusively on observations. Physical and chemical models, instruments, and "mimetic experiments" were common ways of testing for and controlling the illusions of the senses. Scientists switched willy-nilly from observations to experimental studies. But despite their use of models and cross-checks, debates about what precisely counted as an experiment dominated scientific inquiry. The experimental system consisting of a stimulus, a subject, and a recording device—with at least a tenth of a second elapsed from the beginning to the end of this operation—was the site of numerous contestations.

In addition to reevaluating common conceptions of experiment, attention to the tenth of a second can help us rethink what we mean by observation. There is a "context of discovery" opening up every tenth of a second affecting both experiment, observation, and even the most standard measurement techniques.[28]

But let us look even deeper into the history of the tenth of a second. This view leads us to propose alternatives to the epistemological approaches described so far, logical positivist, anti-logical positivist, and post-positivist—all of which, either by accepting it, rejecting it, or reassessing it, continue to be marked by the ghost of positivism. The view (logical positivist, anti-logical positivist, and post-positivist) that scientific production arises from observation, theorization, and experimentation can only be maintained by ignoring essential aspects of the history of the tenth of a second. Scientists observe, develop theory, and experiment, but *first* they redraw metaphysical boundaries, redefining the very meaning of what is considered to be human, nonhuman, and beast. They shape the contours and alter hierarchies between the social, the political, the historical, and the natural. It is more urgent to work on how these divisions arise and to understand their repercussions than it is to study derivative scientific controversies and results.

Instead of trying to fit scientific practices into a semiotic theory of meaning, we can see how traditional concepts of reference were already inoperative in nineteenth-century scientific practices and continued to be disputed

28. I am referring to the famous distinction of "context of discovery" and "context of justification" employed by Hans Reichenbach. For the implications of changing experiment's place from the "context of justification" to the "context of discovery," see Hans-Jörg Rheinberger, *Toward a History of Epistemic Things: Synthesizing Proteins in the Test Tube*, ed. Timothy Lenoir and Hans Gumbrecht, Writing Science (Stanford, Calif.: Stanford University Press, 1997), 15–16.

in certain schools of twentieth-century philosophy. The standard model of understanding our involvement with the world in terms of stimulus, transmission, central processing, reaction, and response already depended on a certain conception of the tenth of a second. To understand scientific practices, then, requires developing new theories of knowledge production grounded on new concepts of signification. The tenth of a second can help us rethink these in broader terms, such as in terms of the relation of movement (marked by difference) to form (marked by repetition).[29] Most importantly, it can help us recover alternative philosophies of science devised during this period.

A focus on the tenth of a second furthers our understanding of the sensorimotor organization of science. New arrangements between keys, bodies, wires, clocks, texts, images, and screens were set in place in attempts to understand this moment and solve the problems it posed. New experiments (such as those based on stimuli, subjects, and recording devices) started to occupy an important place within this organization. Scientists all over the world altered their experimental systems to avoid the effects of reaction time and the tenth-of-a-second errors associated with it. These systems involved new behaviors drawn from practices of spectatorship; new feelings; new intrabody distinctions (such as between retinas, nerves, brains, muscles, and fingers); and even new divisions between matter and life.

Instead of limiting our analysis of science to debates among social groups, we can recapture controversies taking place across wider territories. We can study debates that did not only concern measurement standards, but that concerned the practices of measurement themselves. We can focus on recalcitrant problems that continued—even after consensus on standards and on measurement techniques was reached. Even when these no longer appear at the level of social groups, or individuals, they remain clearly present at the level of different body parts.

By narrowing even more on our object of study (to an object even more basic than experiment) we can further expand our understanding of measurement beyond its social context by seeing just what happens in tenth-of-a-second moments, when scientists all over the world tried to determine

---

29. A number of scholars have noted how a concern with difference and repetition has slowly replaced a concern with identity and contradiction: "Difference and repetition have taken the place of the identical and the negative, of identity and contradiction. . . . The primacy of identity, however conceived, defines the world of representation. But modern thought is born of the failure of representation, of the loss of identities. . . . The modern world is one of simulacra." Gilles Deleuze, *Difference and Repetition*, trans. Paul Patton (New York: Columbia University Press, 1994), xix. Cited in Rheinberger, *Toward a History of Epistemic Things: Synthesizing Proteins in the Test Tube*, 79.

the size of a ruler, the diameter of a planet, the time when a star passed behind the wires of a telescope, the moment of apparent contact between two celestial bodies, or the time when a burst of light appeared and disappeared behind a toothed wheel. This perspective takes us across disciplines, geographies, and even social groups. These low-level activities are only unproblematic if idealized in the manner of positivist and later historians.

Science and Discourse

One of the most important hallmarks of modernity is that this time period was characterized by the emergence of ways of knowing different and superior to discourse. The solution of tenth-of-a second problems was often considered a proof that knowledge could be freed from the vagaries of discourse. The president of the Anthropological Society of Bombay expressed this idea succinctly. He believed cranial measurements were sturdy forms of knowledge because they could be freed from the personal equation: "No one can realize how difficult it is to secure a full and accurate statement on any given point by verbal enquiry from Orientals and still more, from semi savages. . . . Cranial measurements, on the other hand, are probably almost absolutely *free from the personal equation* of the observer."[30] Measurement-based science appeared superior to verbal communication precisely because it could be freed from the personal equation. It was only then able to trump "verbal enquiry."

In the humorous story by Voltaire that I mentioned above, the savants returning from the expedition to measure the shape of the earth agreed about measurement, and not about much else: "We measure lines, we combine numbers, we agree upon two or three things that we understand and argue over the two or three thousand that we do not.'"[31] In a more serious tone, Aldous Huxley, one of the keenest observers of modernity, explained how reaching agreement on matters involving numbers (including cases involving money) was infinitely simpler than reaching consensus on other matters: "You can always discuss figures, haggle over prices, ask a hundred and accept eighty-five. But you cannot discuss hatred, nor haggle over

---

30. Denzil Ibbetson, "The Study of Anthropology in India," *Journal of the Anthropological Society of Bombay* 2 (1890): 121. Cited in Christopher Pinney, "The Parallel Histories of Anthropology and Photography," in *Anthropology and Photography, 1860–1920*, ed. Elizabeth Edwards (New Haven, Conn.: Yale University Press, 1992).
31. Voltaire, "Micromégas: A Philosophical Story," 32–33.

contradictory vanities and prejudices, nor ask for blood and accept a soft answer."[32]

For Habermas, one of the staunchest defenders of modernity, consensus resulted from combining reason with unhampered flows of communication. Measurement practices were ideal exemplars of this process. They were considered a unique form of knowledge superior to others, an "ideal speech situation," which stood out especially in comparison to common discursive practices.[33] Measurement's *inarticulable* qualities gave it much of its strength. Yet for all of its virtues, the modern conception of justice *qua* measurement remains a stunning reminder of just how scarce agreement is. This formulation is extremely limiting.

Foregrounded by the silent, inarticulable structures of consensus exemplified by scientific practices, disagreement appears vocal, vicious, irreconcilable, irrational, and pervasive. But can we understand disagreement by reexamining the forms of agreement against which it is frequently contrasted? Reevaluating these forms of (dis)agreement requires a reexamination of the history of measurement. How can we think about agreement and disagreement in a different, nonmodern, way? Can we conceive of it in a way in which it is no longer related to modern conceptions of measurement? A history of the tenth of a second, drawing much inspiration from critiques of the cinematographic method and new media philosophy, can point toward a new understanding of social agreement and its relation to science. We can also start to reevaluate much longer historical periods.

Longue Durée History

What can we learn through a detailed history of such a short period of time? This moment sheds light not only on the historical period in which it gained importance (from the mid-nineteenth to early twentieth century), but it also helps us understand common historical conceptions. In particular, it offers us a way to rethink "modernity," both as a chronologically delimited period and as a conceptually defined category.

In *La Pensée sauvage* (1962) Claude Lévi-Strauss explained how the scale of time chosen by historians reflected their cultural status. According to the famous anthropologist, civilized cultures organized history by using both

---

32. Aldous Huxley, *Beyond the Mexique Bay* (New York: Harpers and Brothers, 1934), 72.

33. Belonging to "analytic-empirical" discourse, Jürgen Habermas separated the "ideal speech situation" used to describe measurement from other types of communication. For Habermas on science, see Gordon R. Mitchell, "Did Habermas Cede Nature to the Positivists?" *Philosophy and Rhetoric* 36, no. 1 (2003).

"ordinal numbers" to denote succession and "cardinal numbers" to express a distance between them. These numbers fell into distinct classes: "the date 1685 belongs to a class of which 1610, 1648 and 1715 are likewise members, but it means nothing in relation to the class composed of the dates: 1st, 2nd, 3rd, 4th millennium, nor does it mean anything in relation to the class of dates: 23 January, 17 August, 30 September, etc."[34] Comparing history written at the "scale of millennia" against a completely different genre, "hourly history of each day," Lévi-Strauss argued that these two cases belonged to "different species." The smaller the time period between the dates, the more "low-power" these narratives became (approaching biographical genres), and the closer they appeared to the myths of uncivilized cultures.[35] While civilized culture turned to history, "peoples without history" turned to myths. If, finally, in those instances where numbers were rejected entirely and a synchronic view adopted, myth replaced history: "There is no history without dates."[36]

If we place an account of the tenth of a second within Lévi-Strauss's description of historical cultures, it would be even more low powered than biographical genres based on a day and very close to the myths of uncivilized cultures that reject numbers completely. Yet there is also an aspect of the tenth of a second that can help us reevaluate the longer period of time known as modernity and problematize the progression described by Lévi-Strauss. Consequently, we can rethink the great divides underlying many of these discourses, such as between the civilized and the savage, the modern and the primitive, and myth and history. What role has the tenth of a second had in constituting modernity and its divides? How do science and technology uphold these divisions? What fault lines remain below them?

Despite the fact that measurement is perhaps the most original feature of modernity, we have neglected to understand just how it works. By looking at its history through the perspective of the tenth of a second it no longer appears as a result of a linear, progressive development of instruments and techniques.[37] It is also not the product of an analogous development emerg-

34. Claude Lévi-Strauss, *The Savage Mind*, ed. Julian Pitt-Rivers and Ernest Gellner, Nature of Human Societies (Chicago: University of Chicago Press, 1966), 259. First published as Claude Lévi-Strauss, *La Pensée sauvage* (1962).
35. Lévi-Strauss, *The Savage Mind*, 260.
36. Ibid., 258.
37. Examples where quantification was portrayed as a straightforward incremental improvement of instruments and standards include John Heilbron, *Elements of Early Modern Physics* (Berkeley: University of California Press, 1982), 65–67; Charles Coulston Gillispie, *The Edge of Objectivity* (Princeton, N.J.: Princeton University Press, 1960). For a critique of this perspective, see Mary Terrall, "Representing the Earth's Shape: The Polemics Surrounding Maupertuis's Expedition to Lapland," *Isis* 83 (1992): 1.

ing from specific social practices. It is, on the contrary, at the center of great metaphysical debates, even those calling into question divisions between physical and psychical domains, matter and life.

Some recent historians of science have argued that we have never been entirely modern.[38] Part of this assertion is based on a new conception of how science works. While scientists claim to deal solely with nature, a close study of laboratory practices reveals how their work involves a number of hybrids (humans and nonhumans, nature and culture) that only appear separate when scientific results emerge as a finished product. By looking at the history of measurement in a comparable perspective, we can also come to the retrospective realization that we have never measured, at least not in the sense that considers measurement as different and superior to common discursive practices. Instead, we can see how measurement is not a one-directional applied activity but a process of *becoming* that involves the measurer.[39]

Longue durée narratives are often based on common (mis)understandings about the short periods of time that constitute them. If we look more deeply into a history of these moments, all the way into the details pertaining to the determination of the tenth of a second, science no longer appears as the unique territory against which nonmodern spaces can be compared. Debates surrounding the tenth of a second reveal a space where the great dichotomies of modernity no longer hold: the spiritual is mixed in with the mechanical, the nonhuman with the human, the personal with the impersonal, the individual with the social, the natural with the political, and the primitive with the modern.

While problems of the order of a tenth of a second had opened up dangerous floodgates between astronomy, physics, physiology, psychology, art, and philosophy, the gates were soon closed again. In the twentieth century, standard systems of recording were set in place; the relation between analysis and synthesis was settled; technologies once shared by media and science broke off to merge into one or the other but not into both; a separation of the life sciences and physical sciences strengthened; physics emerged as the preeminent science of precision; boundaries among scientists and between scientists and their instruments were reestablished; deviations

38. Bruno Latour, *We Have Never Been Modern* (Cambridge, Mass.: Harvard University Press, 1991).
39. This characteristic of measurement is described by Isabelle Stengers: "In this case, measure and becoming are combined, for the term *measure* does not designate the thing without also designating what becomes capable of measuring it, what the created link with the thing provokes in its ethical, aesthetic, practical and ethological singularity." Isabelle Stengers, *The Invention of Modern Science*, ed. Sandra Buckley, Michael Hardt, and Brian Massumi, trans. Daniel M. Smith, Theory Out of Bounds (Minneapolis: University of Minnesota Press, 2000), 163.

due to individual differences were minimized; a standard account of the expansion of quantification into the human sciences was canonized. These changes reflected the new configuration of science that characterized the twentieth century, when the concern with tenth-of-a-second individual variations was finally put to rest—*but only for a moment.*

POSTSCRIPT

Despite the best efforts of numerous scientists and philosophers, debates pertaining to the tenth of a second did not end in total closure. While some of these debates permitted me to write the preceding conclusion, the work of others has revealed to me that this critical project remains, and will perhaps always remain, open.

In the 1930s, when many solutions to tenth-of-a-second limitations of vision and reaction were widely implemented in the sciences, the social theorist and critic Walter Benjamin referenced this value. In his famous "The Work of Art in the Age of Mechanical Reproduction," he wrote about the tenth of a second in relation to film. Film had turned the "tenth of second" into explosive "dynamite" capable of bursting open the prison-world of modernity. At the moment when the modern world "appeared to have us locked up hopelessly . . . came film and burst this prison-world asunder by *the dynamite of the tenth of a second,* so that now, in the midst of its far-flung ruins and debris, we calmly and adventurously go traveling."[1]

1. Walter Benjamin, "The Work of Art in the Age of Mechanical Reproduction," in *Illuminations*, ed. Hannah Arendt (New York: Schocken, 1969). In the epilogue to his famous text Benjamin's initial optimism waned as he confronted the surging forces of fascism. The complete citation follows: "Our taverns and our metropolitan streets, our offices and furnished rooms, our railroad stations and our factories

Benjamin's famous reference to the tenth of a second formed part of his larger work on film. He hoped that, because of its unique ability to focus on short periods of time, film could reveal an "optical unconscious" analogous to the unconscious of the mind, similarly analyzable, and similarly treatable.[2] In small fractions of a second, the "space informed by human consciousness gives way to a space informed by the unconscious." Studying this "optical unconscious" was not a simple task: "we have no idea at all what happens in the fraction of time when a person *steps out*."[3] But cinematographic instruments could reveal this secret space.

Obtaining detailed information on short moments of time could have an explosive effect on knowledge. Benjamin was especially intrigued by the potentially liberating effects that this knowledge could have on society. Although he was initially optimistic about gaining this knowledge through film, he soon acknowledged that the task was more difficult.

Benjamin reminded us of the mismatch between general knowledge and that arrived at through instruments, such as photographic and cinematographic ones. He used the example of a person walking: "Even if one has a general knowledge of the way people walk, one knows nothing of a person's posture during the fractional second of a stride."[4] What is particularly interesting about Benjamin's views of the tenth of a second is that an instrument's revelation of this moment of time did not occur through a simple expansion of time. It did not represent a straightforward, ever-increasing ability to peer into shorter and shorter periods. It was not necessarily knowledge that was more "precise." Just like the "enlargement of a snapshot does not simply render more precise what in any case was visible" so the magnification of these moments did not simply increase our knowledge of them. This magnification, instead, serve to disclose an entirely "different nature."[5]

With Benjamin, I have used the tenth of a second to imagine what this "different nature" might look like. As he did, I have used it to interrogate a much longer time period—modernity—and to look for alternatives.

---

appeared to have us locked up hopelessly. Then came the film and burst this prison-world asunder by the dynamite of the tenth of a second, so that now, in the midst of its far-flung ruins and debris, we calmly and adventurously go traveling."

2. Ibid.

3. Walter Benjamin, "A Small History of Photography," in *One-Way Street and Other Writings* (London: NLB, 1979 [1931]), 243. Translated as "the fraction of a second when the step is taken" in Walter Benjamin, "A Short History of Photography," in *Classic Essays on Photography*, ed. Alan Trachtenberg (New Haven, Conn.: Leete's Island Books, 1980 [1931]), 203.

4. Benjamin, "The Work of Art in the Age of Mechanical Reproduction."

5. Ibid.

BIBLIOGRAPHY

Archives

Académie de médecine
Archives de l'Académie des sciences (AAS)
Archives du Collège de France
Archives nationales (AN)
Archives Observatoire de Paris (AOP)
Bibliothèque centrale du Muséum national d'histoire naturelle
Bibliothèque de l'Institut de France
Bibliothèque de la Sorbonne
Bibliothèque nationale de France
Bibliothèque Victor Cousin
British Library Manuscript Collection
Einstein Archives (EA)
Harvard College Observatory
Houghton Library
Preußische Staatsbibliothek zu Berlin

Collections of Instruments and Visual Materials

Conservatoire national des arts et métiers
Musée Marey
Musée national Jean-Jacques Henner
Observatoire de Paris
South Kensington Museum in London

Published Sources

"Académie de médecine. Séance du 9 juillet 1878.—Présidence de M. Richet." *Gazette médicale de Paris* 7 (1878).

"Académie des sciences de Paris.—4 décembre 1871." *Revue scientifique* 1 (1871): 574–75.

"Académie des sciences de Paris.—4 décembre 1871." *Revue scientifique* 8 (1871): 574–75.

"Académie des sciences de Paris: Séance du 4 décembre 1871." *Revue scientifique* 8, no. 24 (9 December 1871): 575.

"Académie des sciences de Paris: Séance du 29 juin 1874." *Revue scientifique* 14, no. 1 (4 July 1874): 23.

"Académie des sciences.—6 juillet 1874." *Revue scientifique* 4 (1874): 45.

"Académie des sciences: Séance du 6 juillet." *Le Moniteur scientifique* 16 (1874): 774–75.

"Académie des sciences: Séance du lundi 1er février." *Les Mondes: Revue hebdomadaire des sciences et de leurs applications aux arts et à l'industrie par M. l'abbé Moigno* 19 (1869): 212–13.

"Académie des sciences: Séance du lundi 1er mars." *Les Mondes: Revue hebdomadaire des sciences et de leurs applications aux arts et à l'industrie par M. l'abbé Moigno* 19 (1869): 361.

"Académie des sciences: Séance du lundi 8 février." *Les Mondes: Revue hebdomadaire des sciences et de leurs applications aux arts et à l'industrie par M. l'abbé Moigno* 19 (1869): 231–32.

"Académie des sciences: Séance du lundi 25 janvier." *Les Mondes: Revue hebdomadaire des sciences et de leurs applications aux arts et à l'industrie par M. l'abbé Moigno* 19 (1869): 174.

"Accusés de réception: Archives néerlandaises des sciences exactes et naturelles." *Les Mondes: Revue hebdomadaire des sciences et de leurs applications aux arts et à l'industrie par M. l'abbé Moigno* 20 (1869): 485–86.

Acloque, Paul. "Hippolyte Fizeau et le mouvement de la Terre: Une Tentative méconnue." *La Vie des sciences* 1, no. 2 (1984): 145–58.

Adorno, Theodor. *In Search of Wagner.* Translated by Rodney Livingstone. London: New Left Books, 1981.

Agamben, Giorgio. "*Mysterium disiunctionis.*" In *The Open: Man and Animal,* edited by Werner Hamacher, 13–16. Stanford, Calif.: Stanford University Press, 2004.

Alder, Ken. *Engineering the Revolution: Arms and Enlightenment in France, 1763–1815.* Princeton, N.J.: Princeton University Press, 1997.

———. *The Measure of All Things: The Seven-Year Odyssey and Hidden Error That Transformed the World.* New York: Free Press, 2002.

Andrade, Jules. *Le Mouvement: Mesures de l'étendue et mesures du temps.* Paris: Félix Alcan, 1911.

Angot, Alfred. "Bulletin des sociétés savantes, Académie des sciences de Paris séance du 5 juin 1876: Les images photographiques obtenues au foyer des lunette astronomiques." *Revue scientifique* 17 (17 June 1876): 597.

———. "Bulletin des sociétés savantes, Académie des sciences de Paris séance du 22 mai 1876: Les photographies obtenues au foyer des lunettes astronomiques." *Revue scientifique* 17 (3 June 1876):549.

Arago, François. *Astronomie populaire, oeuvre posthume.* Edited by J.-A. Barral. Vol. 4. Paris: Gide, 1857.

———. "Avertissement des éditeurs." In *Oeuvres complètes,* edited by J.-A. Barral. Paris: Gide et J. Baudry, 1854.

———. "Chapitre XIII: Mesure de la vitesse de la lumière par des observations faites sur la terre à de courtes distances." In *Astronomie populaire, oeuvre posthume,* edited by J.-A. Barral, 418–25. Paris: Gide, 1857.

———. "Le Tonnerre." In *Oeuvres complètes,* edited by J.-A. Barral. Paris: Gide et J. Baudry, 1854.

———. "Mémoire sur les cercles répétiteurs." In *Oeuvres de François Arago,* edited by J.-A. Barral, 115–37. Paris: Théodore Morgand, 1865.

———. "Note sur un moyen très-simple de s'affranchir des erreurs personnelles dans les observations des passages au méridien." *Comptes rendus des séances de l'Académie des sciences* 36 (1853): 276–84.

———. "Notice scientifique sur le tonnerre." *Annuaire pour l'an 1838 par le Bureau des longitudes.* Paris: Bachelier, 1837.

———. "Rapport sur deux mémoires présentés, l'un par M. Eugène Bouvard, l'autre par M. Victor Mauvais, relatifs à l'obliquité de l'écliptique." *Comptes rendus des séances de l'Académie des sciences* 15 (1842): 944–47.

———. "Sur les observations des longitudes et des latitudes géodésiques." In *Oeuvres de François Arago,* edited by J.-A. Barral, 140–48. Paris: Théodore Morgand, 1865.

Arato, Andrew, and Eike Gebhardt, eds. *The Essential Frankfurt School Reader.* New York: Continuum, 1982.

"Archives de physiologie normale et pathologique, M. A. Bloch. Expériences sur la vitesse du courant nerveux sensitif de l'homme." *Revue philosophique de la France et de l'étranger, dirigée par Th. Ribot* 1 (1876): 215–16.

Ash, Mitchell G. "The Self-Presentation of a Discipline: History of Psychology in the United States between Pedagogy and Scholarship." In *Functions and Uses of Disciplinary Histories,* edited by Loren Graham, Wolf Lepenies, and Peter Weingart. Dordrecht: D. Reidel Publishing Company, 1983.

Babbage, Charles. *Reflections on the Decline of Science in England, and Some of its Causes.* New York: A. M. Kelley, 1830. Reprint 1970.

Bache, Meade. "Reaction Time with Reference to Race." *Psychology Review* 6 (1895): 475–83.

Bachelard, Gaston. *The Dialectic of Duration* (1936). Translated by Mary McAllester Jones. Manchester: Clinamen Press, 2000.

Baily, Francis. "Report on the New Standard Scale of this Society. Drawn Up at the Request of the Council, by F. Baily, Esq. F. R. S. and C., and One of the Vice-Presidents of the Society. Presented December 11, 1835." *Memoirs of the Royal Astronomical Society* 9 (1836): 35–184.

Banet-Rivet, P. "La Représentation du mouvement et de la vie." *Revue des deux mondes* 40 (1 August 1907): 590–621.

Baudelaire, Charles. *The Painter of Modern Life.* Translated by Jonathan Mayne. London: Phaidon, 1964.

Bazin, André. "The Myth of Total Cinema" (1946). In *What Is Cinema?* edited by Hugh Gray, 17–22. Berkeley: University of California Press, 2005.

Beaunis, Henri. *L'École du service de santé militaire de Strasbourg et la Faculté de médecine de Strasbourg de 1856–1870.* Nancy: Berger-Levrault et Cie, 1888.

———. *Nouveaux Éléments de physiologie humaine comprenant les principes de la physiologie comparée et de la physiologie générale.* Paris: J.-B. Baillière et fils, 1876.

———. *Nouveaux Éléments de physiologie humaine comprenant les principes de la physiologie comparée et de la physiologie générale.* 2nd ed. Vol. 2. Paris: J.-B. Baillière et fils, 1881.

———. *Nouveaux Éléments de physiologie humaine comprenant les principes de la physiologie comparée et de la physiologie générale.* 3rd ed. Vol. 1. Paris: J.-B. Baillière et fils, 1888.

———. *Nouveaux Éléments de physiologie humaine comprenant les principes de la physiologie comparée et de la physiologie générale.* 3rd ed. Vol. 2. Paris: J.-B. Baillière et fils, 1888.

————. "Société de psychologie physiologique: I.—Influence de la durée de l'expectation sur le temps de réaction des sensations visuelles." *Revue philosophique de la France et de l'étranger, dirigée par Th. Ribot* 20 (1885): 330–33.

————. "Sur la comparaison du temps de réaction des différentes sensations." *Revue philosophique de la France et de l'étranger, dirigée par Th. Ribot* 15 (1883): 611–20.

————. "Sur le temps de réaction des sensations olfactives." *Comptes rendus des séances de l'Académie des sciences* 96 (1883): 387.

Becquerel, Jean. "Critique par M. J. Becquerel de l'ouvrage Durée et simultanéité de M. Bergson." *Bulletin scientifique des étudiants de Paris* (March 1923).

————. "Préface." In *La Relativité: Exposé sans formules des théories d'Einstein et réfutation des erreurs contenues dans les ouvrages les plus notoires*, v–xviii. Paris: Etienne Chiron, 1923.

Bellynck, R. P. A. "Correspondance des Mondes: R. P. A. Bellynck, à Namur." *Les Mondes: Revue hebdomadaire des sciences et de leurs applications aux arts et à l'industrie par M. l'abbé Moigno* 19 (1869): 558–59.

Benjamin, Walter. "A Short History of Photography" (1931). In *Classic Essays on Photography,* edited by Alan Trachtenberg, 199–216. New Haven, Conn.: Leete's Island Books, 1980.

————. "A Small History of Photography" (1931). In *One-Way Street and Other Writings,* 240–57. London: NLB, 1979.

————. "The Work of Art in the Age of Mechanical Reproduction." In *Illuminations,* edited by Hannah Arendt, 217–51. New York: Schocken, 1969.

Benoît, J.-René. "De la précision dans la détermination des longueurs en métrologie." In *Rapports présentés au congrès international de physique réuni à Paris en 1900,* edited by Ch.-Éd. Guillaume and L. Poincaré. Paris: Gauthier-Villars, 1900.

Benrubi, Isaak. *Souvenirs sur Henri Bergson.* Neuchâtel: Delachaux & Niestlé, 1942.

Bergson, Henri. "Bergson à E. Peillaube." *Revue de philosophie* 24 (July 1924): 440.

————. *Correspondances.* Edited by André Robinet. Paris: Presses Universitaires de France, 2002.

————. *Creative Evolution.* Mineola, N.Y.: Dover, 1998.

————. "Discussion avec Einstein." In *Mélanges,* 1340–47. Paris: Presses Universitaires de France, 1972.

————. "Durée et simultanéité: À propos de la théorie d'Einstein" (1922). In *Mélanges,* 58–244. Paris: Presses Universitaires de France, 1972.

————. "La Pensée et le mouvant" (1934). In *Oeuvres*, 1251–1482. Paris: Presses Universitaires de France, 1991.

————. "Le Parallélisme psycho-physique et la métaphysique positive" (1901). In *Mélanges*, 463–502. Paris: Presses Universitaires de France, 1972.

————. "Les Temps fictifs et les temps réel" (1924). In *Mélanges*, 1432–49. Paris: Presses Universitaires de France, 1972.

————. "Les Temps fictifs et les temps réel." *Revue de philosophie* 24, no. 3 (1924): 241–60.

————. *Mélanges*. Paris: Presses Universitaires de France, 1972.

————. "Préface" (1923). In *Mélanges*, 59–61. Paris: Presses Universitaires de France, 1972.

————. *Two Sources of Morality and Religion.* Translated by R. Ashley Audra and Claudesley Brereton. New York: Henry Holt and Company, 1935.

Berman, Louis. *The Personal Equation.* New York: Century Co., 1925.

Bernard, Claude. *Principes de médecine expérimentale.* Paris: Presses Universitaires de France, 1947.

Bessel, Wilhelm Friedrich. "Personal Gleichung." *Astronomische Beobachtungen* 8 (1822).

Biagioli, Mario, and Peter Galison, eds. *Scientific Authorship: Credit and Intellectual Property in Science.* New York: Routledge, 2003.

Biezunski, Michel. *Einstein à Paris: Le Temps n'est plus . . .* Vincennes: Presses Universitaires de Vincennes, 1991.

Bigg, Charlotte. "Photography and the Labour History of Astrometry: The Carte du Ciel." In *The Role of Visual Representations in Astronomy,* edited by K. Hentschel and A. Wittman, 90–106. Thun: Deutsch, 2000.

Bigourdan, Guillaume. "Notice sur la vie et travaux de M. Ch. Wolf." *Comptes rendus des séances de l'Académie des sciences* 157 (1918): 46–48.

Binet, Alfred. "Travaux du laboratoire de psychologie physiologique (Hautes Études): La Perception de la durée dans les réactions simples." *Revue philosophique de la France et de l'étranger, dirigée par Th. Ribot* 33 (1892): 650–59.

Binet, Léon. "A. d'Arsonval, physiologiste et médecin." In *Cérémonie commémorative en l'honneur du centenaire de la naissance du professeur Arsène d'Arsonval 1851–1940,* edited by L'Union des associations scientifiques et industrielles françaises. Paris: Draeger, 1952.

Bloch, A.-M. "Caractères différentiels des sensations électriques et tactiles." *Travaux du laboratoire de M. Marey* 3 (1877): 123–36.

————. "Deuxième note relative aux expériences de M. Grigorescu." *Comptes rendus des séances et mémoires de la Société de biologie* 3 (1891): 477–78.

————. "Durée de la persistance des sensations de tact dans les différentes régions du corps." *Travaux du laboratoire de M. Marey* 4 (1880): 259–61.

————. "Expériences nouvelles sur la vitesse du courant nerveux sensitif chez l'homme." *Comptes rendus des séances et mémoires de la Société de biologie* 1 (1884): 343–44.

————. "Expériences nouvelles sur la vitesse du courant nerveux sensitif chez l'homme." *Journal de l'anatomie et de la physiologie normales et pathologiques de l'homme et des animaux* 20 (1884): 284–90.

————. "Expériences sur la vitesse du courant nerveux sensitif de l'homme." *Archives de physiologie normale et pathologique, publiées par MM. Brown-Séquard, Charcot, Vulpian* 2 (1875): 588–623.

————. "Expériences sur la vitesse relative des transmissions visuelles, auditives, et tactiles." *Journal de l'anatomie et de la physiologie normales et pathologiques de l'homme et des animaux* 20 (1884): 1–37.

————. "Note à propos de la communication faite par M. G. Gregorescu [sic], le 16 mai." *Comptes rendus des séances et mémoires de la Société de biologie* 3 (1891): 379.

————. "Psychologie: La Vitesse comparative des sensations." *Revue scientifique* 39 (7 May 1887): 585–89.

Blumenthal, Arthur L. "A Reappraisal of Wilhelm Wundt." *American Psychologist* 30 (1975): 1081–86.

Bois-Reymond, Emil du. "Vitesse de la transmission de la volonté et de la sensation à travers les nerfs: Conférence de M. du Bois-Reymond à l'Institution Royale de la Grande-Bretagne." *Revue scientifique* 4, no. 3 (15 December 1866): 33–41.

Boothby, Richard. *Freud as Philosopher: Metapsychology after Lacan.* New York: Routledge, 2001.

Boring, Edwin G. "The Beginning and Growth of Measurement in Psychology." In *Quantification: A History of the Meaning of Measurement in the Natural and Social Sciences,* edited by Harry Woolf, 108–18. Indianapolis: Bobbs-Merrill, 1961.

————. *A History of Experimental Psychology.* New York: Appleton-Century-Crofts, 1929.

The Born-Einstein Letters: Friendship, Politics and Physics in Uncertain Times. New York: Macmillan, 2005.

Boynton, Percy H. *Some Contemporary Americans: The Personal Equation in Literature.* Chicago: University of Chicago Press, 1924.

Brain, Robert M. "The Graphic Method: Inscription, Visualization and Measurement in Nineteenth-Century Science and Culture." Ph. D. thesis, University of California, Los Angeles, 1996.

Brain, Robert M., and M. Norton Wise. "Muscles and Engines: Indicator Diagrams and Helmholtz's Graphical Methods." In *The Science Studies Reader,* ed. Mario Biagioli. New York: Routledge Press, 1999.

Braun, Marta. *Picturing Time: The Work of Etienne-Jules Marey (1830–1904).* Chicago: University of Chicago Press, 1992.

Brentano, Franz. *Psychologie du point de vue empirique.* Edited by L. Lavelle and R. Le Senne. Paris: Aubier, 1944.

Brooks, G. P., and R. C. Brooks. "The Improbable Progenitor." *Journal of the Royal Astronomical Society of Canada* 73 (1979): 9–23.

Brooks III, John I. "Philosophy and Psychology at the Sorbonne, 1885–1913." *Journal of the History of the Behavioral Sciences* 29 (April 1993): 123–45.

Brown-Séquard Charles-Édouard. "Bibliographie: Recherches expérimentales sur la durée des processus psychiques les plus simples et sur la vitesse des courants nerveux à l'état normale et à l'état pathologique, par le Dr. A. Rémond (de Metz)." *Archives de physiologie normal et pathologique, publiées par MM. Brown-Séquard, Charcot, Vulpian* 1 (1889): 612–14.

———. "Recherches sur la transmission des impressions de tact, de chatouillement, de douleur, de température et de contraction (sens musculaire)." *Journal de la physiologie de l'homme et des animaux, publié sous la direction du docteur Brown-Séquard* 6 (1863): 124–45, 232–48, 581–646.

Brunier, Serge, and Jean-Pierre Luminet. *Cosmos: Du Romantisme à l'avant-garde.* Paris: Gallimard, 1999.

———. *Éclipses, les rendez-vous célestes.* Paris: Bordas, 1999.

———. *Trajectoires du rêve.* Paris: Paris-Musées/Actes Sud, 2003.

Buccola, Gabriele. *La legge del tempo nei fenomeni del pensiero: Saggio di psicologia sperimentale.* Biblioteca Scientifica Internazionale, Vol. 37. Milano: Fratelli Dumolard, 1883.

Callandreau, Octave. "Histoire abrégée des déterminations de la parallaxe solaire." *Revue scientifique* 2 (1881): 39–43.

Camus, Jean, and Henri Nepper. "Les Réactions psychomotrices et émotives des trépanés." *Paris médical* (3 June 1916).

———. "Mesure des réactions psychomotrices des candidats à l'aviation." *Paris médical* (18 March 1916).

———. "Temps de réactions psychomotrices des candidats à l'aviation." *Comptes rendus des séances de l'Académie des sciences* 163 (1916): 106–7.

Canales, Jimena. "Einstein, Bergson, and the Experiment That Failed: Intellectual Cooperation at the League of Nations." *Modern Language Notes* 120, no. 5 (2005): 1168–91.

————. "Exit the Frog, Enter the Human: Physiology and Experimental Psychology in Nineteenth-Century Astronomy." *British Journal for the History of Science* 34 (June 2001): 173–97.

————. "Movement before Cinematography: The High-Speed Qualities of Sentiment." *Journal of Visual Culture* 5, no. 3 (2006): 275–94.

————. "The Single Eye: Re-evaluating *Ancien Régime* Science." *History of Science* 39 (March 2001): 71–94.

Carroy, Jacqueline. "Théodule Ribot et la naissance d'une psychologie scientifique." In *L'Anhédonie: Le Non-Plaisir et la psychopathologie,* edited by M.-L. Bourgeois. Paris: Masson, 1999.

Carroy, Jacqueline, and Régine Plas. "The Origins of French Experimental Psychology: Experiment and Experimentalism." *History of the Human Sciences* 9, no. 1 (1996): 73–84.

Chambre des députés. "Chambre des députés—Annexe no. 445: Séance du 3 août 1876." *Journal officiel de la République française* (14 September 1876): 695–97.

Chapman, A. "The Transits of Venus." *Endeavour* 22, no. 4 (1998).

Charpentier, Augustin. "Recherches sur la vitesse des réactions d'origine rétinienne." *Archives de physiologie normale et pathologique, publiées par MM. Brown-Séquard, Charcot, Vulpian* 15 (1883): 599–635.

Charpentier, Thomas Victor. "Th. Ribot.—La psychologie allemande contemporaine." *Revue philosophique de la France et de l'étranger, dirigée par Th. Ribot* 9 (1880): 350–57.

Chauvois, Louis. *D'Arsonval: Soixante-cinq ans à travers la science.* Paris: J. Oliven, 1937.

"Chronique." *L'Année psychologique* 22 (1920–21): 598.

Clark, William, Jan Golinski, and Simon Schaffer. "Introduction." In *The Sciences in Enlightened Europe,* edited by William Clark, Jan Golinski, and Simon Schaffer, 3–31. Chicago: University of Chicago Press, 1999.

Clerke, Agnes. *A Popular History of Astronomy during the Nineteenth Century.* Edinburgh: Adam and Charles Black, 1885.

Cohen, I. Bernard. "Commentary." In *Scientific Change: Historical Studies in the Intellectual, Social and Technical Conditions for Scientific Discovery and Technical Invention, from Antiquity to the Present,* edited by A. C. Crombie, 466–71. London: Heinemann, 1963.

Coissac, Michel G. *Histoire du cinématographe.* Paris: Éditions du "Cinéopse," 1925.

Colin, Gabriel. *Traité de physiologie comparée des animaux considérée dans les rapports avec les sciences naturelles, la médecine, la zootechnie et l'économie rurale.* 3rd ed. Vol. 1. Paris: J.-B. Baillière, 1886.

————. *Traité de physiologie comparée des animaux domestiques.* Vol. 1. Paris: J.-B. Baillière, 1854.

Colladon, Daniel. *Comptes rendus des séances de l'Académie des sciences* 109 (1889): 12–15.

Collins, Harry, and Trevor Pinch. *The Golem: What You Should Know about Science.* Cambridge: Cambridge University Press, 1998.

Commission internationale de coopération intellectuelle. In *Mélanges,* 1352–53. Paris: Presses Universitaires de France, 1922.

"Communications 1. Sur l'importance au point de vue médical des signes extérieurs des fonctions de la vie." *Bulletin de l'Académie de médecine* 7 (1878): 610–26.

"Compte rendu des académies. Bulletin de l'Académie des sciences. Séance du lundi 31 octobre 1859." *Journal du progrès des sciences médicales et de l'hydrothérapie rationelle* 4 (1859): 390.

Comptes rendus des séances de l'Académie des sciences 68 (1 March 1869): 525.

Corbin, Brenda G. "Étienne Léopold Trouvelot, 19th Century Artist and Astronomer." *Bulletin of the American Astronomical Society* 34 (2002).

Cornu, Alfred. *Détermination de la vitesse de la lumière d'après des expériences exécutées en 1874 entre l'observatoire et Montlhéry, extrait des Annales de l'Observatoire de Paris (Mémoires, tome XIII).* Paris: Gauthier-Villars, 1876.

————. "Détermination nouvelle de la vitesse de la lumière." *Comptes rendus des séances de l'Académie des sciences* 76 (1873): 338–42.

————. "Détermination nouvelle de la vitesse de la lumière." *Journal de l'École polytechnique* 27 (1874): 133–80.

————. "La Décimalisation de l'heure et de la circonférence." *L'Éclairage électrique* 11 (1897): 385–90.

Cosmos: Du Romantisme à l'Avant-garde. Paris: Gallimard, 1999.

Costabel, Pierre. "L. Foucault et H. Fizeau: Exploitation d'une information nouvelle." *La Vie des sciences* 1, no. 3 (1984): 235–49.

Crary, Jonathan. *Suspensions of Perception: Attention, Spectacle, and Modern Culture.* Cambridge, Mass.: MIT Press, 1999.

Crosland, Maurice. "Popular Science and the Arts: Challenges to Cultural Authority in France during the Second Empire." *British Journal for the History of Science* 34 (2001): 301–22.

Cuyer, Edouard. *Allures du cheval démontrées à l'aide d'une planche coloriée, découpée, superposée et articulée.* Paris: J.-B. Baillière et fils, 1883.

Cyre, Le. "Dépouillement des journaux étrangers, par nos traducteurs volontaires: Durée de la transmission des sensations." *Les Mondes: Revue heb-*

domadaire des sciences et de leurs applications aux arts et à l'industrie par M. l'abbé Moigno 19 (1869): 374–76.

Danziger, Kurt. *Constructing the Subject: Historical Origins of Psychological Research.* Edited by William R. Woodward and Mitchell G. Ash, Cambridge Studies in the History of Psychology. Cambridge: Cambridge University Press, 1990.

Darius, Jon. *Beyond Vision: One Hundred Historic Scientific Photographs.* Oxford: Oxford University Press, 1984.

D'Arsonval, Arsène. "Chronomètre à embrayage magnétique pour la mesure directe des phénomènes de courte durée (de une seconde à 1/500e de seconde)." *Comptes rendus des séances et mémoires de la Société de biologie* 3 (1886): 235–36.

Daston, Lorraine, and Peter Galison. "The Image of Objectivity." *Representations* 40 (1992): 81–128.

———. *Objectivity.* New York: Zone Books, 2007.

Daumas, Maurice. "Precision of Measurement and Physical and Chemical Research in the Eighteenth Century." In *Scientific Change: Historical Studies in the Intellectual, Social and Technical Conditions for Scientific Discovery and Technical Invention, from Antiquity to the Present,* edited by A. C. Crombie, 418–30. London: Heinemann, 1963.

Dauriac, Lionel. "Le Mouvement philosophique: De la psychologie expérimentale en Allemagne." *Revue politique et littéraire* 17 (1879): 241–47.

Davanne. "Rapport sur la XIe exposition de la Société française de photographie." *Bulletin de la Société Française de Photographie* 22 (1876): 225–34.

Debru, Claude. "Ernst Mach et la psychophysiologie du temps." *Philosophia Scientiae* 7, no. 2 (2003): 59–91.

de Jaager, Johan Jacob. "Introduction." In *Origins of Psychometry: Johan Jacob de Jaager, Student of F. C. Donders on Reaction Time and Mental Processes,* edited by Josef Brozek and Maarten S. Sibinga. Nieuwkoop: B. de Graaf, 1970.

———. "Reaction Time and Mental Processes" (1865). In *Origins of Psychometry: Johan Jacob de Jaager student of F. C. Donders on Reaction Time and Mental Processes,* edited by Josef Brozek and Maarten S. Sibinga. Nieuwkoop: B. de Graaf, 1970.

Delaunay, Charles. "Notice sur la distance du Soleil à la Terre, extrait de l'Annuaire pour l'an 1866, publié par le Bureau des longitudes." In *Recueil de mémoires, rapports et documents relatifs à l'observation du passage de Vénus sur le Soleil,* 3–110. Paris: Firmin Didot, 1874.

Delboeuf, Joseph. "Étude psychophysique: Recherches théoriques et expérimentales." *Mémoires couronnés et autres mémoires publiés par l'Académie royale des sciences, des lettres et des beaux-arts de Belgique* 23 (1873): 1–116.

———. "La loi psychophysique et le nouveau livre de Fechner." *Revue philosophique de la France et de l'étranger, dirigée par Th. Ribot* 5 (1878): 34–63, 127–55.

———. "La Mesure des sensations: Réponses à propos du logarithme des sensations." *Revue scientifique* 8 (1875): 1014–17.

Deleuze, Gilles. *Cinema 1: The Movement-Image.* Translated by Hugh Tomlinson and Barbara Habberjam. Minneapolis: University of Minnesota Press, 1986.

———. *Difference and Repetition.* Translated by Paul Patton. New York: Columbia University Press, 1994.

———. "Hélène Cixous ou l'écriture stroboscopique" (1972). In *L'Île déserte et autres textes: Textes et entretiens 1953–1974,* edited by David Lapoujade, 320–22. Paris: Les Éditions de Minuit, 2002.

Delhoume, Léon. *De Claude Bernard à D'Arsonval.* Paris: J.-B. Baillière & fils, 1939.

DeVorkin, David H. *Henry Norris Russell: Dean of American Astronomers.* Princeton, N.J.: Princeton University Press, 2000.

———. "Venus 1882: Public, Parallax, and HNR." *Sky and Telescope* 22, no. 4 (1982).

Dewey, John. "The Reflex Arc Concept in Psychology." *Psychological Review* 3 (1896): 357–70.

Diamond, Solomon. "Wilhelm Wundt." In *Dictionary of Scientific Biography,* 526–29. New York, 1970.

Dick, Steven J., Wayne Orchiston, and Tom Love. "Simon Newcomb, William Harkness, and the Nineteenth-Century American Transit of Venus Expeditions." *Journal for the History of Astronomy* 29 (August 1998): 221–55.

Documents relatifs aux mesures des épreuves photographiques. Recueil de mémoires, rapports et documents relatifs à l'observation du passage de Vénus sur le Soleil, extrait du tome III, vol. 3. Paris: Gauthier-Villars, 1882.

Donders, F. C. "Deux instruments pour la mesure du temps nécessaire pour les actes psychiques (Extrait des Archives Néerlandaises)." *Journal de l'anatomie et de la physiologie normales et pathologiques de l'homme et des animaux* 5 (1868): 676–78.

———. "La Vitesse des actes psychiques." *Archives Néerlandaises* 3 (1868): 269–317.

———. "On the Speed of Mental Processes." In *Attention and Performance II: Proceedings of the Donders Centenary Symposium on Reaction Time,* edited by W. G. Koster, 412–31. Amsterdam: North Holland, 1969.

Douglass, Paul. "Bergson and Cinema: Friends or Foes." In *The New Bergson,* edited by John Mullarkey, 209–27. Manchester: Manchester University Press, 1999.

Dubois, Edmond. *Les Passages de Vénus sur le disque solaire.* Paris: Gauthier-Villars, 1873.

———. "Nouvelle méthode pour déteminer la parallaxe de Vénus sans attendre les passages de 1874 ou 1882." *Comptes rendus des séances de l'Académie des sciences* 69 (20 December 1869): 1290.

Duhem, Pierre. *The Aim and Structure of Physical Theory.* Translated by Philip P. Wiener. New York: Atheneum, 1962.

Duncombe, Raynor L. "Personal Equation in Astronomy." *Popular Astronomy* 53 (1945): 3–13, 63–76.

Duruy, Victor. *Journal officiel de la République française* (2 February 1869).

Eder, J.-M. *La Photographie instantanée, son application aux arts et aux sciences.* Translated by O. Campo. Translated from the 2nd German ed. Paris: Gauthier-Villars et fils, 1888.

Editor. "Study of Types of Character." *Mind* 2, no. 8 (1877): 573–74.

Einstein, Albert. "How I Became a Zionist." In *The Collected Papers of Albert Einstein,* 234–37. Princeton, N.J.: Princeton University Press, 2002.

———. "How I Became a Zionist." *Jüdische Rundschau* (1921): 351–52.

———. *Ideas and Opinions.* New York: Three Rivers, 1984.

———. *Letters to Solovine.* New York: Philosophical Library, 1987.

———. *Mein Weltbild.* Amsterdam: Querido, 1934.

Exner, Sigmund. "Experimentelle Untersuchung der einfachsten psychischen Processe. Erste Abhandlung: Die persönliche Gleichung." *Archiv für die gesammte Physiologie des Menschen und der Thiere, herausgegeben von Dr. E. F. W. Pflüger* 7 (1873): 601–60.

———. "Experimentelle Untersuchung der einfachsten psychischen Processe. Zweite Abhandlung: Ueber Reflexzeit und Rückenmarksleitung." *Archiv für die gesammte Physiologie des Menschen und der Thiere, herausgegeben von Dr. E. F. W. Pflüger* 8 (1874): 526–37.

Fabre, Ch. *Les Industries photographiques.* Encyclopédie industrielle. Paris: Gauthier-Villars.

———. *Renseignements photographiques.* Matériaux pour l'histoire primitive de l'homme. Paris: Ch. Reinwald et cie, 1878.

Faye, Hervé. "Académie des sciences: Séance du lundi 4 janvier." *Les Mondes: Revue hebdomadaire des sciences et de leurs applications aux arts et à l'industrie par M. l'abbé Moigno* 19 (1869): 42–44.

———. "Académie des sciences: Séance du lundi 11 janvier." *Les Mondes: Revue hebdomadaire des sciences et de leurs applications aux arts et à l'industrie par M. l'abbé Moigno* 19 (1869): 85–86.

———. "Association française pour l'avancement des sciences, congrès de Lille, conférences publiques: Le Prochain Passage de Vénus sur le Soleil." *Revue scientifique* 14, no. 16 (17 October 1874): 361–69.

———. *Cours d'astronomie de l'École polytechnique.* Vol. 1. Paris: Gauthier-Villars, 1881.

———. *Cours d'astronomie et de géodésie de l'École polytechnique.* Paris: Gauthier-Villars, 1883.

———. "Note sur les nouvelles tables des planètes intérieures." *Comptes rendus des séances de l'Académie des sciences* 54 (1862): 630.

———. "Rapport sur le rôle de la photographie dans l'observation du passage de Vénus, extrait des comptes rendus hebdomadaires des séances de l'Académie des sciences, séance du 2 septembre 1872." In *Recueil de mémoires, rapports et documents relatifs à l'observation du passage de Vénus sur le Soleil,* 227–36. Paris: Firmin Didot frères, fils, et Cie, 1874.

———. "Sur l'art de pointer et ses conditions physiologiques." *Comptes rendus des séances de l'Académie des sciences* 71 (19 December 1870): 872–75.

———. "Sur la méthode des coïncidences appliquée à la mesure de la vitesse du son et sur la détermination des longitudes." *Comptes rendus des séances de l'Académie des sciences* 55 (1862): 521–23.

———. "Sur les erreurs d'origine physiologique." *Comptes rendus des séances de l'Académie des sciences* 59 (12 September 1864): 473–80.

———. "Sur les observations du Soleil." *Comptes rendus des séances de l'Académie des sciences* 28 (1849): 241–44.

———. "Sur les passages de Vénus et la parallaxe du Soleil." *Comptes rendus des séances de l'Académie des sciences* 68 (4 January 1869): 42–50.

———. "Sur les passages de Vénus et la parallaxe du Soleil." *Comptes rendus des séances de l'Académie des sciences* 68 (11 January 1869): 69–74.

———. "Sur les photographies de l'éclipse du 15 mars, présentées par MM. Porro et Quinet," *Comptes rendus des séances de l'Académie des sciences* 46 (1858): 705–10.

———. "Sur l'expédition de M. Janssen." *Comptes rendus des séances de l'Académie des sciences* 71 (12 December 1870): 819–22.

———. "Sur l'observation photographique des passages de Vénus et sur un appareil de M. Laussedat." *Comptes rendus des séances de l'Académie des sciences* 70 (14 March 1870): 541–48.

Féré, Charles. "Le Travail et le temps de réaction." *Comptes rendus des séances et mémoires de la Société de biologie* 4 (1892): 432–53.

———. "Note sur la physiologie de l'attention." *Comptes rendus des séances et mémoires de la Société de biologie* 2 (1890): 484–88.

———. "Note sur le temps d'association, sur les conditions qui le font varier et sur quelques conséquences de ses variations." *Comptes rendus des séances et mémoires de la Société de biologie* 2 (1890): 173–80.

———. "Observations faites sur les épileptiques à l'aide du sphygmomètre de M. Bloch." *Comptes rendus des séances et mémoires de la Société de biologie* 5 (1888).

Fizeau, Armand. *Annales de chimie et de physique* 2 (1864).

Flammarion, Camille. "Le Passage de Vénus: Résultats des expéditions françaises." *La Nature: Revue des sciences et de leurs applications aux arts et à l'industrie* 3 (1875): 356–58.

———. "Le Prochain Passage de Vénus et la mesure des distances inaccessibles." *La Nature: Revue des sciences et de leurs applications aux arts et à l'industrie* 2 (1874): 386–91.

———. "Les Progrès de l'astronomie pendant l'année 1887." *L'Astronomie* 7 (1888): 161–73.

Fonvielle, Wilfrid de. *L'Astronomie moderne.* Paris: Germer Baillière, 1868.

———. *Le Mètre international définitif.* Paris: Masson, 1875.

Forbes, George. *The Transit of Venus,* Nature Series. London: Macmillan, 1874.

Fossé, David. "Étienne-Léopold Trouvelot: Peintre du firmament." *Ciel et espace,* no. 421 (2005).

Foucault, Léon. "Détermination expérimentale de la vitesse de la lumière; description des appareils." *Comptes rendus des séances de l'Académie des sciences* 55 (1862): 792–96.

———. *Mesure de la vitesse de la lumière. Étude optique des surfaces: Mémoires de Léon Foucault.* Edited by H. Abraham, H. Gautier, H. Le Chatelier, and J. Lemoine, Les Classiques de la science. Paris: Armand Colin, 1913.

———. "Méthode générale pour mesurer la vitesse de la lumière dans l'air et les milieux transparents. Vitesses relatives de la lumière dans l'air et dans l'eau. Projet d'expérience sur la vitesse de propagation du calorique rayonnant." *Comptes rendus des séances de l'Académie des sciences* 30 (1850): 551–60.

————. "Sur les vitesses relatives de la lumière dans l'air et dans l'eau." *Annales de chimie et physique* 41 (1854): 129–64.

Foucault, Michel. *The Archaeology of Knowledge.* Translated by A. M. Sheridan Smith. New York: Pantheon, 1972.

————. *Discipline and Punish: The Birth of the Prison.* Translated by Alan Sheridan. 2nd Vintage Books ed. New York: Vintage Books, 1995.

————. *Power/Knowledge: Selected Interviews and Other Writings, 1972–1977.* Translated by Colin Gordon, Leo Marshall, John Mepham, and Kate Soper. Edited by Colin Gordon. New York: Pantheon Books, 1977.

————. "What Is Enlightenment?" In *The Foucault Reader,* edited by Paul Rabinow, 32–50. New York: Pantheon, 1984.

Fraisse, Paul. "The Evolution of Experimental Psychology." In *History and Method,* ed. Jean Piaget, Paul Fraisse, and Maurice Reuchlin, vol. 1 of *Experimental Psychology: Its Scope and Method,* ed. Jean Piaget, Paul Fraisse, and Maurice Reuchlin. New York: Basic Books, 1968.

Frängsmyr, Tore, J. L. Heilbron, and Robin R. Rider, eds. *The Quantifying Spirit in the 18th Century.* Berkeley: University of California Press, 1990.

Freud, Sigmund. *Psychopathology of Everyday Life.* London: T. Fisher Unwin, 1901.

Galison, Peter. "Einstein's Clock: The Place of Time." *Critical Inquiry* 26 (Winter 2000): 355–89.

————. *Einstein's Clocks, Poincaré's Maps: Empires of Time.* New York: W. W. Norton and Company, 2003.

————. *Image and Logic: A Material Culture of Microphysics.* Chicago: University of Chicago Press, 1997.

————. "Judgment against Objectivity." In *Picturing Science, Producing Art,* edited by Caroline A. Jones and Peter Galison, 327–59. New York: Routledge, 1998.

Galison, Peter, and Caroline A. Jones. "Trajectories of Production: Laboratories/Factories/Studios." In *Laboratorium,* edited by Hans Ulrich Obrist and Barbara Vanderlinden, 205–10. Cologne: Dumont, 2001.

Galton, Francis. "Address." *Nature* 16 (1877): 344–47.

Garver, M. M. "Art. XXIV.—The Periodic Character of Voluntary Nervous Action." *American Journal of Science and Arts* 20 (1880): 189–93.

Gautier, A. "Travaux scientifiques étrangers: Travaux et faits astronomiques récents." *Revue scientifique* 8 (22 December 1871): 616–22.

Gautier, Émile. "Monsieur Trouvelot." *La Science française* 5, no. 14 (3 May 1895).

———. "Un Martyr de la science." *La Science française,* no. 19 (7 June 1895).

Gavarret. "Observations à l'occasion du procès-verbal. III. Méthode graphique." *Bulletin de l'Académie de médecine* 7 (1878): 759–69.

Genovesi, Angelo. "Henry Bergson: Lettere a Einstein." *Filosofia* 49, no. 1 (1998): 3–39.

Gillispie, Charles Coulston. *The Edge of Objectivity.* Princeton, N.J.: Princeton University Press, 1960.

Gley, Eugène. "Notices bibliographiques." *Revue philosophique de la France et de l'étranger, dirigée par Th. Ribot* 15 (1883): 566–67.

Goldstein, Jan. "The Hysteria Diagnosis and the Politics of Anticlericalism in Late Nineteenth-Century France." *Journal of Modern History* 54 (1982): 209–39.

———. "Mutations of the Self in Old Regime and Postrevolutionary France: From Ame to Moi and to Le Moi." In *Biographies of Scientific Objects,* edited by Lorraine Daston, 86–116. Chicago: University of Chicago Press, 2000.

Gonnessiat, François. *Recherches sur l'équation personnelle dans les observations astronomiques de passage.* Annales de l'Université de Lyon. Vol. 3. Paris: G. Masson, 1892.

Gooday, Graeme J. N. *The Morals of Measurement: Accuracy, Irony and Trust in Late Victorian Electrical Practice.* Cambridge: Cambridge University Press, 2004.

Gough, J. B. "Armand-Hippolyte-Louis Fizeau." In *Dictionary of Scientific Biography,* edited by Charles Coulston Gillispie, 18–21. New York: Scribner's, 1970–86.

Gouhier, Henri. "Avant-Propos." In *Mélanges,* vi–ixxiii. Paris: Presses Universitaires de France, 1972.

Grigorescu, G. "Application du chronomètre électrique de M. d'Arsonval au diagnostic des myélites." *Comptes rendus des séances et mémoires de la Société de biologie* 3 (1891): 364–65.

———. "Réponse à la note de M. Bloch, relative à ma communication du 16 mai." *Comptes rendus des séances et mémoires de la Société de biologie* 3 (1891): 475–78.

Grimaud, Octave. "L'examen psycho-physiologique des soldats mitrailleurs." *La Science et la vie* 13, no. 36 (1917–18): 121–28.

Guérard, Albert. *Personal Equation.* New York: W. W. Norton and Company, 1948.

Guéroult, Georges. "Sur un moyen d'emmagasiner les gestes et les jeux de physionomie." *Comptes rendus des séances de l'Académie des sciences* 108 (1889): 1030.

————. "Sur une application nouvelle de la photographie et du phénakis-tiscope." *Comptes rendus des séances de l'Académie des sciences* 122 (1896): 404–6.

Guillaume, Ch.-Éd. "L'Oeuvre du Bureau international des poids et mesures." In *La création du Bureau international des poids et mesures et son oeuvre,* edited by Ch.-Éd. Guillaume, 33–258. Paris: Gauthier-Villars, 1927.

Guillaume, Ch.-Éd., and Lucien Poincaré. "Avertissement." In *Rapports présentés au congrès international de physique,* edited by Ch.-Éd. Guillaume and Lucien Poincaré, v–xv. Paris: Gauthier-Villars, 1900.

Gunter, P. A. Y., ed. *Bergson and the Evolution of Physics.* Knoxville, Tenn.: University of Tennessee Press, 1969.

Habermas, Jürgen. "Modernity versus Postmodernity." In *The Continental Aesthetics Reader,* edited by Clive Cazeaux, 268–77. London: Routledge, 2000.

Hacking, Ian. "Biopower and the Avalanche of Numbers." *Humanities and Society* 5 (1983): 279–95.

Harding, Sandra G., ed. *Can Theories be Refuted? Essays on the Duhem-Quine Thesis.* Dordrecht, Netherlands: Reidel, 1976.

Heidegger, Martin. *Being and Time* (1927). New York: Harper-Collins, 1962.

Heilbron, John. *Elements of Early Modern Physics.* Berkeley: University of California Press, 1982.

Helmholtz, Hermann von. "On the Methods of Measuring Very Small Portions of Time, and Their Application to Physiological Purposes." *The London, Edinburgh and Dublin Philosophical Magazine and Journal of Science* 4 (1853): 313–25.

————. "Ueber die Methoden, kleinste Zeittheile zu messen, und ihre Anwendung für physiologische Zwecke." *Königsberger Naturwissenschaftliche Unterhaltungen* 2 (1851): 169–89.

Herman, Jan K., and Brenda G. Corbin. "Trouvelot: From Moths to Mars." *Sky and Telescope* 72 (1986): 566–68.

Hingley, Peter, and Françoise Launay. "Passages de Vénus, 1874 et 1882." In *Dans le champ des étoiles: Les Photographes et le ciel, 1850–2000.* Paris: Réunion des musées nationaux, 2000.

Hirsch, Adolph. "Expériences chronoscopiques sur la vitesse des différentes sensations et de la transmission nerveuse." *Bulletin de la Société des sciences naturelles de Neuchâtel (1861 à 1864)* 6 (1864): 100–14.

————. "Sur les erreurs personelles: Lettre de M. A. Hirsch, directeur de l'Observatoire cantonal de Neuchâtel, à M. R. Radau." *Le Moniteur scientifique: Journal des sciences pures et appliquées* 8 (1866): 315–17.

Hoffleit, E. Dorrit. "Trouvelot, Étienne Léopold." In *Dictionary of Scientific Biography,* ed. Charles Coulston Gillispie. New York: Scribner's, 1970.

Hoffmann, Christoph. "'Helmholtz' Apparatuses: Telegraphy as a Working Model of Nerve Physiology." *Philosophie Scientiae* 7 (2003): 129–49.

———. *Unter Beobachtung: Naturforschung in der Zeit der Sinnesapparate.* Göttingen: Wallstein Verlag, 2006.

Holmberg, Gustav. "Mechanizing the Astronomer's Vision: On the Role of Photography in Swedish Astronomy, c. 1880–1914." *Annals of Science* 53 (1996): 609-16.

Holmes, Frederic L., and Kathryn M. Olesko. "The Images of Precision: Helmholtz and the Graphical Method in Physiology." In *The Values of Precision,* ed. M. Norton Wise. Princeton, N.J.: Princeton University Press, 1995.

Huxley, Aldous. *Beyond the Mexique Bay.* New York: Harpers and Brothers, 1934.

Ibbetson, Denzil. "The Study of Anthropology in India." *Journal of the Anthropological Society of Bombay* 2 (1890): 117–46.

Isaachsen, D. "Introduction historique." In *La Création du Bureau international des poids et mesures et son oeuvre,* 1–31. Paris: Gauthier-Villars, 1927.

James, William. *Principles of Psychology.* Vol. 1. New York: Dover Publications, 1950.

———. "Report on Mrs. Piper's Hodgson-Control." In *William James on Psychical Research,* edited by Gardner Murphy and Robert O. Ballou, 115–210. Clifton, N.J.: Augustus M. Kelley, 1973.

Jamin, Jules. "Correspondance." *Comptes rendus des séances de l'Académie des sciences* 99 (1884): 959.

Janet, Pierre. "La Psychologie expérimentale et comparée." In *Le Collège de France (1530–1930), Livre jubilaire composé à l'occasion de son quatrième centenaire,* 223–34. Paris: Presses Universitaires de France, 1930.

Janiczek, Paul M. "Remarks on the Transit of Venus Expedition of 1874." In *Sky with Ocean Joined: Proceedings of the Sesquicentennial Symposia of the U.S. Naval Observatory, December 5 and 8,* ed. Steven J. Dick and LeRoy E. Dogget. Washington, D.C.: U.S. Naval Observatory, 1983.

Janiczek, Paul M., and L. Houchins. "Transits of Venus and the American Expedition of 1874." *Sky and Telescope* 48 (1974).

Janssen, Jules. "Bulletin des sociétés savantes, Académie des sciences de Paris.—2 Octobre 1876: Observations sur les passages devant le Soleil de corps intra-mercuriels." *Revue scientifique* 18 (14 October 1876): 383.

———. "De la nécessité actuelle de la topographie: L'Exemple des Boers." In *Lectures académiques: Discours,* 295–98. Paris: Hachette et cie, 1903.

———. "Discours prononcé à la première séance du congrès international de photographie, tenu à Paris, au palais Trocadéro, le 6 août 1889, par M. Janssen, président du comité d'organisation du congrès." In

Procès-verbaux et résolutions, du congrès international de photographie tenu à Paris du 6 au 17 août 1889, Paris: Imprimerie Nationale, 1889.

———. "En l'honneur de la photographie: Discours prononcé au banquet annuel de la Société française de la photographie, juin 1888." In *Oeuvres scientifiques recueillies et publiées par Henri Dehérain,* 86–90. Paris: Société d'éditions géographiques, maritimes et coloniales, 1930.

———. "La Photographie céleste." *Revue scientifique* (14 January 1888): 33.

———. *La Photographie céleste, conférence faite au congrès de Toulouse.* Paris: Administration de deux revues, 1888.

———. "Les Méthodes en astronomie physique: Discours prononcé comme président du congrès le 26 août 1882." In *Lectures académiques: Discours,* 211–21. Paris: Hachette et cie, 1903.

———. "Méthode pour obtenir photographiquement les circonstances physiques des contacts avec les temps correspondants, communication faite à la Commission du passage de Vénus, dans sa séance du 15 février 1873." In *Recueil de mémoires, rapports et documents relatifs à l'observation du passage de Vénus sur le Soleil,* 295–98. Paris: Firmin Didot, 1874.

———. "Note sur l'observation du passage de la planète Vénus sur le Soleil." *Comptes rendus des séances de l'Académie des sciences* 96 (29 January 1883): 288–92.

———. "Passage de Vénus: Méthode pour obtenir photographiquement l'instant des contacts, avec les circonstances physiques qu'ils présentent." *Comptes rendus des séances de l'Académie des sciences* 76 (17 March 1873): 677–79.

———. "Présentation de quelques spécimens de photographies solaires obtenues avec un appareil construit pour la mission du Japon." *Comptes rendus des séances de l'Académie des sciences* 78 (22 June 1874): 1730–31.

———. "Sur la constitution de la surface solaire et sur la photographie envisagée comme moyen de découverte en astronomie physique." *Comptes rendus des séances de l'Académie des sciences* 85 (31 December 1877): 1250–55.

———. "Sur l'éclipse totale du 22 décembre prochain." *Comptes rendus des séances de l'Académie des sciences* 71 (24 October 1870): 531–33.

Jastrow, Joseph. *The Time-Relations of Mental Phenomena.* Fact and Theory Papers. New York: N. D. C. Hodges, 1890.

Jones, H. Spencer. "The Early History of the ICSU 1919–1946." *ICSU Review* 2 (1960): 169–87.

"Journal de l'anatomie et de la physiologie, etc., L. Lalanne. Note sur la durée de la sensation tactile." *Revue philosophique de la France et de l'étranger, dirigée par Th. Ribot* 1 (1876): 650.

Journal officiel de la République française (1 September 1872): 5796.

Jung, Carl G., ed. *Diagnostische Assoziationsstudien: Beiträge zur experimentellen Psychopathologie.* Leipzig: Barth, 1906.

———. "Die psychologische Diagnose des Tatbestandes," *Schweizerische Zeitschrift für Strafrecht* 18 (1905). Republished as Carl G. Jung, "The Psychological Diagnosis of Evidence," in *Experimental Researches: The Collected Works of C. G. Jung.* London: Routledge, 1973.

———. "Editorial" (1933). In *The Collected Works of C. G. Jung,* edited by Herbert Read, Michael Fordham, Gerhard Adler, and William McGuire, 533–34. London: Routledge, 1970.

———. *Gesammelte Werke.* Vol. 10. Zurich: Rascher, 1958–94.

———. "Psychological Types" (1921). In *The Collected Works of C. G. Jung,* vol. 6, ed. Herbert Read, Michael Fordham, Gerhard Adler, and William McGuire. London: Routledge, 1970.

Kant, Immanuel. *Metaphysische Anfangsgründe der Naturwissenschaft.* Leipzig: Johann Friedrich Hartknoch, 1800.

Kennelly, A. E. "The Metric System of Weights and Measures." *Scientific Monthly* 23, no. 6 (1926): 549–51.

Kevles, Daniel J. "'Into Hostile Political Camps': The Reorganization of International Science in World War I." *Isis* 62 (1971): 47–60.

Kittler, Friedrich. *Gramophone, Film, Typewriter.* Edited by Timothy Lenoir and Hans Ulrich Gumbrecht, Writing Science. Stanford, Calif.: Stanford University Press, 1999.

Kries, Johannes von, and Felix Auerbach. "Die Zeitdauer einfachster psychischer Vorgänge." *Archiv für Physiologie: Physiologische Abtheilung des Archives für Anatomie und Physiologie, herausgegeben von Dr. Emil du Bois-Reymond* (1877): 297–378.

Kuhn, Thomas S. "The Function of Measurement in Modern Physical Science." In *Quantification: A History of the Meaning of Measurement in the Natural and Social Sciences,* edited by Harry Woolf, 31–63. Indianapolis: Bobbs-Merrill, 1961.

———. "Mathematical versus Experimental Traditions in the Development of Physical Science." In *The Essential Tension: Selected Studies in Scientific Tradition and Change,* 31–65. Chicago: University of Chicago Press, 1977.

Kusch, Martin. "Recluse, Interlocutor, Interrogator: Natural and Social Order in Turn-of-the-Century Psychological Research Schools." *Isis* 86 (1995): 419–39.

"La Conférence internationale du passage de Vénus." *Revue scientifique* 29 (14 January 1882): 42–46.

La Direction. "Réponse de Henri Bergson." *Revue de philosophie* 24 (July 1924): 440.

Lacassagne, A. *Alphonse Bertillon: l'homme, le savant, la pensée philosophique.* Lyon: A. Rey, 1914.

Lahy, Jean Maurice. *La sélection psychophysiologique des travailleurs.* Paris: Dunod, 1927.

———. *Le système Taylor et la physiologie du travail professionnel.* Paris: Masson, 1916.

———. "Sur la psycho-physiologie du soldat mitrailleur." *Comptes rendus des séances de l'Académie des sciences* 163 (1916): 33–35.

Lalande, André. "Philosophy in France, 1922–1923." *Philosophical Review* 33, no. 6 (1924): 535–59.

Lalanne, Léon. "Note sur la durée de la sensation tactile." *Journal de l'anatomie et de la physiologie normales et pathologiques de l'homme et des animaux* 12 (1876): 449–55.

———. "Sur la durée de la sensation tactile." *Comptes rendus des séances de l'Académie des sciences* 90 (1876): 1314–16.

Lankford, John. "The Impact of Photography on Astronomy." In *Astrophysics and Twentieth-Century Astronomy,* ed. Owen Gingerich, General History of Astronomy. Cambridge: Cambridge University Press, 1984.

———. "Photography and the Nineteenth-Century Transits of Venus." *Technology and Culture* 28, no. 3 (July 1987): 648–57.

"The Late M. Trouvelot." *Nature* 52 (1895): 11.

Latour, Bruno. *Science in Action: How to Follow Scientists and Engineers through Society.* Cambridge, Mass.: Harvard University Press, 1987.

———. *We Have Never Been Modern.* Cambridge, Mass.: Harvard University Press, 1991.

———. "Why Has Critique Run Out of Steam? From Matters of Fact to Matters of Concern." *Critical Inquiry* 30, no. 2 (2004): 225–48.

Latour, Bruno, and Steve Woolgar. *Laboratory Life: The Construction of Scientific Facts.* Princeton, N.J.: Princeton University Press, 1986.

Launay, Françoise. "Jules Janssen et la photographie." In *Dans le champ des étoiles: Les photographes et le ciel 1850–2000,* 22–31. Paris: Réunion des musées nationaux, 2000.

———. "Trouvelot à Meudon: Une 'affaire' et huit pastels." *L'Astronomie* 117 (2003): 453–61.

Le Verrier, Urbain. "Bulletin des sociétés savantes, Académie des sciences de Paris.—2 Octobre 1876: Les planètes intra-mercurielles." *Revue scientifique* 18 (14 October 1876): 383.

———. "Sur les masses des planètes et la parallaxe du soleil." *Comptes rendus des séances de l'Académie des sciences* 75 (22 July 1872): 165–72.

———. "Sur les passages de Vénus et la parallaxe du Soleil." *Comptes rendus des séances de l'Académie des sciences* 68 (4 January 1869): 49–50.

———. "Théorie et tables du mouvement apparent du Soleil." *Annales de l'Observatoire impérial de Paris* 4 (1858): 1–262.

Lévy-Leblond, Jean-Marc. "Le Boulet d'Einstein et les boulettes de Bergson." Edited by Frédéric Worms. *Annales bergsoniennes* 3 (2007): 236–58.

Lévi-Strauss, Claude. *La Pensée sauvage.* Paris: Plon, 1962.

———. *The Savage Mind.* Edited by Julian Pitt-Rivers and Ernest Gellner. The Nature of Human Societies. Chicago: University of Chicago Press, 1966.

Lightman, Bernard. "The Visual Theology of Victorian Popularizers of Science: From Reverent Eye to Chemical Retina." *Isis* 91 (2000): 651–80.

Locke, John. *An Essay Concerning Human Understanding.* London: Penguin Books, 1997.

Luxenberg, Alisa. "'The art of correctly painting the expressive lines of the human face': Duchenne de Boulogne's Photographs of Human Expression and the École des Beaux-Arts." *History of Photography* 25 (2001): 201–12.

Lyttleton, R. A. "History of the Mass of Mercury." *Quarterly Journal of the Royal Astronomical Society* 21 (1980): 400–13.

M. E., "Les Photographies d'éclairs." *La Lumière électrique* 25, no. 37 (10 September 1887): 534–35.

Mach, Ernst. *Contributions to the Analysis of the Sensations* (1886). Translated by C. M. Williams. La Salle, Ill.: Open Court, 1897.

Marey, Etienne-Jules. "Fusil photographique." *Bulletin de la Société française de photographie* 28 (1882): 127–33.

———. "The History of Chronophotography." *Annual Report of the Board of Regents of the Smithsonian Institution, Showing the Operations, Expenditures, and Condition of the Institution for the Year Ending June 30, 1901* (1902): 317–40.

———. "La Méthode graphique dans les sciences expérimentales." *Travaux du laboratoire de M. Marey* 1 (1876): 123–61.

———. *La Méthode graphique dans les sciences expérimentales et principalement en physiologie et en médecine.* Paris: G. Masson, 1878.

———. "La Photochronographie et ses applications à l'analyse des phénomènes physiologiques." *Archives de physiologie normale et pathologique, publiées par MM. Brown-Séquard, Charcot, Vulpian* 1 (1889): 508–612.

———. "Leçon d'ouverture: Vitesse des actes nerveux et cérébraux.—Le vol dans la série animale, Collège de France, histoire naturelle des corps organisés, cours de M. Marey." *Revue scientifique* 6, no. 4 (26 December 1868): 61–64.

———. "Nouvelles modifications du chronophotographe par M. E. J. Marey." In Louis Gastine, *La Chronophotographie, encyclopédie scientifique des aide-mémoire.* Paris: Gauthier-Villars et fils.

Maxwell, James Clerk. *Nature* 21 (1880): 314–15.

McTurnan, Lawrence. *The Personal Equation.* New York: Atkinson, Mentzer & Grover, 1910.

Méry, Gaston. "A l'observatoire de Meudon." *La Libre Parole* (27 May 1895).

———. "L'Observatoire de Meudon." *La Libre Parole* (29 May 1895).

——— [G. M.]. "L'Observatoire de Meudon." *La Libre Parole* (6 June 1895).

———. "Une éclipse de Lune: Histoire de deux arbres qu'on n'a pas voulu abattre et d'un astronome qui en est mort." *La Libre Parole* (24 May 1895).

Metz, André. *La Relativité: Exposé sans formules des théories d'Einstein et réfutation des erreurs contenues dans les ouvrages les plus notoires.* Paris: Etienne Chiron, 1923.

———. *Temps, espace, relativité.* Science et philosophie, Vol. 2. Paris: Gabriel Beauchesne, 1928.

Meunier, Stanislas. "Académie des sciences." *La Nature: Revue des sciences et de leurs applications aux arts et à l'industrie* 3 (1875): 78.

Michelson, Albert A. "Comparison of the International Meter with the Wave Length of the Light of Cadmium." *Astronomy and Astrophysics* 12 (1893): 556–60.

———. *Détermination expérimentale de la valeur du mètre en longueurs d'ondes lumineuses: Travaux et mémoires du Bureau international des poids et mesures.* Paris: Gauthier-Villars et fils, 1894.

———. "Experimental Determination of the Velocity of Light." *Proceedings of the American Association for the Advancement of Science* 28 (1880): 124–60.

———. "On the Application of Interference Methods to Astronomical Measurements." *Philosophical Magazine* 30 (1890): 1–21.

Michelson, Albert A., and Edward Williams Morley. "On the Feasibility of Establishing a Light-wave as the Ultimate Standard of Length." *American Journal of Science* 38, no. 225 (1889): 181–86.

Ministère de l'Instruction publique. "Cinquième séance: Jeudi 13 octobre 1881." In *Conférence internationale du passage de Vénus.* Paris: Imprimerie nationale, 1881.

———. *Conférence internationale du passage de Vénus.* Paris: Imprimerie nationale, 1881.

———. "Deuxième séance: Jeudi 6 octobre 1881." In *Conférence internationale du passage de Vénus*. Paris: Imprimerie nationale, 1881.

———. "Instructions pour l'observation des contacts." In *Conférence internationale du passage de Vénus*, 28–33. Paris: Imprimerie nationale, 1881.

———. "Lettre." *Comptes rendus des séances de l'Académie des sciences* 68 (1 February 1869): 205–6.

———. "Observation du passage de Vénus." *Les Mondes: Revue hebdomadaire des sciences et de leurs applications aux arts et à l'industrie par M. l'abbé Moigno* 19 (18 February 1869): 239–40.

———. "Première séance: Mercredi 5 octobre 1881." In *Conférence internationale du passage de Vénus*, 1–11. Paris: Imprimerie nationale, 1881.

Mitchell, Gordon R. "Did Habermas Cede Nature to the Positivists?" *Philosophy and Rhetoric* 36, no. 1 (2003): 1–21.

Morus, Iwan Rhys. "'The Nervous System of Britain': Space, Time and the Electric Telegraph in the Victorian Age." *British Journal for the History of Science* 33 (2000): 455–75.

Mossé-Bastide, Rose-Marie. *Bergson éducateur*. Paris: Presses Universitaires de France, 1955.

Mouchez, Ernest. "Astronomie." *Comptes rendus des séances de l'Académie des sciences* 87 (1878): 970.

Moussette, Charles. "Orage du 12 mai 1886. La Foudre en spirale." *Comptes rendus des séances de l'Académie des sciences* 103 (1886): 30.

Müller, Johannes. *Elements of Physiology*. 2 vols. Translated by William Baly. London: Taylor and Walton, 1838–42.

———. *Handbuch der Physiologie des Menschen*. Vol. 1. Coblentz: Hölscher, 1834.

Münsterberg, Hugo. *Psychology and Industrial Efficiency*. Boston: Mifflin, 1913.

Murray, Charles. "Afterword." In *The Bell Curve: Intelligence and Class Structure in American Life*, 553–75. New York: Simon & Schuster, 1996.

Murray, Charles, and Richard J. Herrnstein. *The Bell Curve: Intelligence and Class Structure in American Life*. New York: Free Press, 1994.

Nathan, Otto, and Heinz Norden, eds. *Einstein on Peace*. New York: Avenel, 1960.

"Nécrologie: E.-L. Trouvelot." *La Nature: Revue des sciences et de leurs applications aux arts et à l'industrie* 23 (1895): 530.

Nicolas, Serge, and Ludovic Ferrand. "Pierre Janet au Collège de France." *Psychologie et histoire* 1 (2000): 131–50.

Nietzsche, Friederich. *The Will to Power*. Translated by Walter Kaufmann and R. J. Hollingdale. New York: Random House, 1967.

Nordmann, Charles. "Einstein à Paris." *Revue des deux mondes* 8 (1922): 926–37.

Obrist, Hans Ulrich, and Barbara Vanderlinden, eds. *Laboratorium*. Cologne: Dumont, 2001.

"Observations à l'occasion du procès-verbal. II. Méthode graphique." *Bulletin de l'Académie de médecine* 7 (1878): 722–36.

Olesko, Kathryn M. "Error & the Personal Equation." In *Oxford Companion to the History of Modern Science*, edited by J. L. Heilbron, 271–73. Oxford: Oxford University Press, 2003.

———. "The Meaning of Precision: The Exact Sensibility in Early Nineteenth-Century Germany." In *The Values of Precision*, edited by M. Norton Wise, 103–34. Princeton, N.J.: Princeton University Press, 1995.

Olesko, Kathryn M., and Frederic L. Holmes. "Experiment, Quantification, and Discovery: Helmholtz's Early Physiological Researches, 1843–50." In *Hermann von Helmholtz and the Foundations of Nineteenth-Century Science*, ed. David Cahan. Berkeley: University of California Press, 1993.

Olivier, Louis. "Physiologie: La Photographie du mouvement." *Revue scientifique* 30 (1882): 802–11.

Ostwald, Wilhelm. *Les Grands Hommes*. Translated by Marcel Dufour. Bibliothèque de philosophie scientifique. Paris: Ernest Flammarion, 1912.

Pais, Abraham. *Einstein Lived Here*. Oxford: Clarendon Press, 1994.

Pang, Alex Soojung-Kim. *Empire and the Sun: Victorian Solar Eclipse Expeditions*. Stanford, Calif.: Stanford University Press, 2002.

———. " 'Stars Should Henceforth Register Themselves': Astrophotography at the Early Lick Observatory." *British Journal for the History of Science* 30 (1997).

———. "Technology, Aesthetics, and the Development of Astrophotography at the Lick Observatory." In *Inscribing Science*, ed. Timothy Lenoir. Stanford, Calif.: Stanford University Press, 1998.

———. "Victorian Observing Practices, Printing Technologies, and Representations of the Solar Corona, 1: The 1860's and 1870's." *Journal for the History of Astronomy* 25 (1994).

———. "Victorian Observing Practices, Printing Technologies, and Representations of the Solar Corona, 2: The Age of Photomechanical Reproduction." *Journal for the History of Astronomy* 26 (1995).

Parrot, Françoise. "La Psychologie scientifique française et ses instruments au début du XXe siècle." In *Studies in the History of Scientific Instruments*, edited by Christine Blondel, Françoise Parot, Anthony Turner, and Mari Williams. London: Rogers Turner Books Ltd, 1989.

Pearson, Karl. "On the Mathematical Theory of Errors of Judgment with Special Reference to the Personal Equation." In *Early Statistical Papers*, 377–441. Cambridge: Cambridge University Press, 1948.

———. "On the Mathematical Theory of Errors of Judgment with Special Reference to the Personal Equation." *Philosophical Transactions of the Royal Society of London* 198 (1902): 235–99.

Peck, Harry Thurston. *The Personal Equation.* New York: Harpers, 1897.

Peters, C. A. F. *Ueber die Bestimmung des Längenunterschiedes zwischen Altona und Schwerin, ausgeführt im Jahre 1858 durch galvanische Signale.* Altona: Hammerich & Lesser, 1861.

"Physiologie du système nerveux." *Journal du progrès des sciences médicales et de l'hydrothérapie rationelle* 4 (1859): 323–24.

Picavet, Fr. "Philosophes français contemporains, M. Théodule Ribot." *Revue politique et littéraire* 2, no. 18 (1894): 590–95.

Pinney, Christopher. "The Parallel Histories of Anthropology and Photography." In *Anthropology and Photography, 1860–1920,* edited by Elizabeth Edwards. New Haven, Conn.: Yale University Press, 1992.

Planck, Max. *Eight Lectures on Theoretical Physics.* Translated by A. P. Willis. Mineola, N.Y.: Dover, 1998.

Polanyi, Michael. *Personal Knowledge.* Chicago: University of Chicago Press, 1958.

Popper, Karl R. *The Logic of Scientific Discovery.* London: Routledge, 1959.

———. *The Open Society and Its Enemies.* Princeton, N.J.: Princeton University Press, 1950.

Pouchet, Georges. "Le Système nerveux et l'intelligence." *Revue des deux mondes* 93 (15 June 1871): 717–48.

Prinz, W. "Étude de la structure des éclairs par la photographie." *La Lumière électrique* 30, no. 42 (20 October 1888): 126–30.

"Psychological Literature: L'Énergie et la vitesse des mouvements volontaires. Ch. Féré. Revue philosophique. Juillet, 1889." *American Journal of Psychology* 2 (1889): 647–48.

"Psychological Literature: Recherches expérimentales sur la durée des actes psychiques les plus simples et sur la vitesse des courants nerveux à l'état pathologique. A. Rémond. Paris: Octave Doin. 1888. pp. 135." *American Journal of Psychology* 2 (1888): 486–87.

Puiseux, Victor. "Note sur la détermination de la parallaxe du Soleil, par l'observation du passage de Vénus sur cet astre en 1874." *Comptes rendus des séances de l'Académie des sciences* 68 (8 February 1869): 321–24.

Rabinbach, Anson. *The Human Motor: Energy, Fatigue and the Origins of Modernity.* Berkeley: University of California Press, 1990.

Radau, Rodolphe. *L'Acoustique ou les phénomènes du son.* Bibliothèque des merveilles. Paris: L. Hachette et Cie, 1867.

————. "La Vitesse de la volonté." *Le Moniteur scientifique* 5 (1868): 88–92.

————. "Le Passage de Vénus du 9 décembre 1874." *Le Moniteur scientifique* 16 (1874): 474–84.

————. "Le Passage de Vénus du 9 décembre 1874." *Revue des deux mondes* 1 (15 January 1874): 434–49.

————. "Les Applications scientifiques de la photographie." *Revue des deux mondes* 25 (15 February 1878): 872–90.

————. "Sur les erreurs personnelles." *Le Moniteur scientifique* 7 (1865): 977–85, 1025–32.

————. "Sur les erreurs personnelles." *Le Moniteur scientifique* 8 (1866): 97–102, 155–61, 207–17.

————. "Ueber die persönliche Gleichungen." *Repertorium für physikalische Technik für mathematische und astronomische Instrumentenkunde, herausgegeben von Dr. Ph. Carl* 2 (1866): 115–56.

————. "Ueber die persönlichen Gleichungen bei Beobachtungen derselben Erscheinungen durch verschiedene Beobachter." *Repertorium für physikalische technik für mathematische und astronomische instrumentenkunde, herausgegeben von Dr. Ph. Carl* 1 (1865): 202–18.

Rakov, Vladmir A., and Martin A. Uman. *Lightning: Physics and Effects.* Cambridge: Cambridge University Press, 2003.

Regnault. "Remarque de M. Regnault à l'occasion de la note de M. Faye." *Comptes rendus des séances de l'Académie des sciences* 59 (1864): 479–80.

Rémond, A. *Recherches expérimentales sur la durée des actes psychiques les plus simples et sur la vitesse des courants nerveux à l'état normal et à l'état pathologique.* Paris: Octave Doin, 1888.

René, Albert. "Étude expérimentale sur la vitesse de transmission nerveuse chez l'homme: Durée d'un acte cérébral et d'un acte réflexe, vitesse sensitive, vitesse motrice." *Gazette des hôpitaux* 55 (1882): 276–77, 283–84, 307–9, 363–64, 373.

Renoliet, Jean-Jacques. *L'UNESCO oubliée: La Société des nations et la coopération intellectuelle (1919–1946).* Paris: Publications de la Sorbonne, 1999.

"Revue des périodiques étrangers: Philosophische Studien, Hgg. von W. Wundt. 1883. 4e Heft." *Revue philosophique de la France et de l'étranger, dirigée par Th. Ribot* 15 (1883): 574–77.

Rheinberger, Hans-Jörg. *Toward a History of Epistemic Things: Synthesizing Proteins in the Test Tube.* Edited by Timothy Lenoir and Hans Gumbrecht. Writing Science. Stanford, Calif.: Stanford University Press 1997.

Ribot, Théodule. "Analyses: G. Buccola. La legge del tempo nei fenomeni del pensiero: saggio di psicologia sperimentale." *Revue philosophique de la France et de l'étranger, dirigée par Th. Ribot* 16 (1883): 420–27.

———. "De la durée des actes psychiques d'après les travaux récents." *Revue philosophique de la France et de l'étranger, dirigée par Th. Ribot* 1 (1876): 267–88.

———. "Histoire des sciences: Leçon d'ouverture du cours de psychologie expérimentale et comparée de Collège de France." *Revue scientifique* 15 (1888): 450–55.

———. *La Psychologie allemande contemporaine (école expérimentale).* Paris: Libraire Germer Baillière et Cie, 1879.

———. "La Psychologie allemande contemporaine: M. Wilhelm Wundt." *Revue scientifique* 8 (1875): 723–32.

———. "La Psychologie allemande contemporaine: M. Wilhelm Wundt." *Revue scientifique* 8 (1875): 751–60.

———. "La Psychologie physiologique en Allemagne." *Revue scientifique* 7 (1874): 553–63.

———. "La Psychologie physiologique en Allemagne: M. W. Wundt." *Revue scientifique* 9 (1875): 505–16.

———. "La Psychologie physiologique en Allemagne: M. W. Wundt." *Revue scientifique* 9 (1875): 544–49.

———. "Psychologie: La Psychologie physiologique en 1889." *Revue scientifique* 18 (1889): 177–80.

Richet, Charles. "Cerveau." In *Dictionnaire de physiologie,* edited by Charles Richet, 1–36. Paris: Alcan, 1898.

———. "De la durée des actes psychiques élémentaires." *Revue philosophique de la France et de l'étranger, dirigée par Th. Ribot* 6 (1878): 393–96.

———. "Du minimum de temps dans la réaction psycho-physiologique aux excitations visuelles et auditives." *Comptes rendus des séances de l'Académie des sciences* 163 (1916): 78–80.

———. "Études sur la vitesse et les modifications de la sensibilité chez les ataxiques: Note lue à la Société de biologie, dans sa séance du 17 Juin 1876." *Comptes rendus des séances et mémoires de la Société de biologie* 3 (1877): 79–89.

———. "L'Avenir de la psychologie." *Revue scientifique* 50 (3 September 1892): 292–96.

———. "L'Unité psychologique du temps." *Comptes rendus des séances de l'Académie des sciences* 173, no. 25 (1921): 1313–17.

———. "Le professeur Charles Richet, autobiographie recueillie par le Dr. Pierre Maurel." In *Les Biographies médicales, notes pour servir à l'histoire*

de la médecine et des grands médecins. Paris: J.-B. Baillière et fils,
1932.

———. *Physiologie des muscles et des nerfs: Leçons professées à la Faculté
de médecine en 1881.* Paris: Germer Baillière et cie, 1882.

———. *Recherches expérimentales et cliniques sur la sensibilité.* Paris: G.
Masson, 1877.

Rodin, Auguste. *L'Art: Entretiens réunis par Paul Gsell.* Paris: Bernard Grasset, 1924.

Roseveare, N. T. "Leverrier to Einstein: A Review of the Mercury Problem."
Vistas in Astronomy 23 (1979): 165–71.

Rothermel, Holly. "Images of the Sun: Warren De La Rue, George Biddell
Airy, and Celestial Photography." *British Journal for the History of Science* 26 (1993).

Rouse, Joseph. "Philosophy of Science and the Persistent Narratives of Modernity." *Studies in History and Philosophy of Science* 22, no. 1 (1991):
141–62.

Sadoul, Georges. *Histoire du cinéma mondial des origines à nos jours.* Paris:
Denoël, 1946.

Sagra, Ramon de la. "Académie des Sciences: Séance du lundi 1er février."
*Les Mondes: Revue hebdomadaire des sciences et de leurs applications
aux arts et à l'industrie par M. l'abbé Moigno* 19 (1869): 213–15.

Sanford, Edmund C. "Personal Equation." *American Journal of Psychology* 2
(1889): 3–38, 271–98, 403–30.

Schaffer, Simon. "Accurate Measurement Is an English Science." In *Values
of Precision,* edited by M. Norton Wise, 135–72. Princeton, N.J.: Princeton University Press, 1995.

———. "Astronomers Mark Time: Discipline and the Personal Equation."
Science in Context 2 (1988): 115–45.

———. "Late Victorian Metrology and Its Instrumentation: A Manufactory
of Ohms." In *Invisible Connections, Instruments, Institutions and Science,*
edited by Robert Bud and Susan Cozzens, 23–56. Bellingham, Wash.:
SPIE Optical Engineering Press, 1992.

———. "A Science Whose Business Is Bursting: Soap Bubbles as Commodities in Classical Physics." In *Things That Talk: Object Lessons from Art
and Science,* edited by Lorraine Daston, 147–92. New York: Zone Books,
2004.

Schmidgen, Henning. "The Donders's Machine." *Configurations* 13 (2005):
211–56.

Schmidgen, Henning, and Jacqueline Carroy. "Reaktionsversuche in
Leipzig, Paris und Würzburg: Die deutsch-französische Geschichte eines

psychologischen Experiments, 1890–1910." *Medizinhistorisches Journal* 39 (2004): 27–55.

Schroeder-Gudehus, Brigitte. "Les Congrès scientifiques et la politique de coopération internationale des académies de science." *Relations internationales* 62 (1990): 135–48.

Schuster, Arthur. *Biographical Fragments.* London: Macmillan, 1932.

Secord, Anne. "Botany on a Plate: Pleasure and the Power of Pictures in Promoting Early Nineteenth-Century Scientific Knowledge." *Isis* 93 (2002): 28–57.

Seelig, Carl, ed. *Albert Einstein: Eine dokumentarische Biographie.* Zürich: Europa Verlag, 1954.

Shamdasani, Sonu. *Jung and the Making of Modern Psychology: The Dream of a Science.* Cambridge: Cambridge University Press, 2003.

Shapin, Steven. *A Social History of Truth: Civility and Science in Seventeenth-Century England.* Edited by David L. Hull. Science and Its Conceptual Foundations. Chicago: University of Chicago Press, 1994.

Shapin, Steven, and Simon Schaffer. *Leviathan and the Air-Pump.* Princeton, N.J.: Princeton University Press, 1985.

Sheehan, William, and Richard Baum. *In Search of Planet Vulcan: The Ghost in Newton's Clockwork Universe.* New York: Plenum, 1997.

Sibum, H. Otto. "Les gestes de la mesure: Joule, les pratiques de la brasserie et la science." *Annales HSS* 4–5 (1998).

Sicard, Monique. "Passage de Vénus: Le Revolver photographique de Jules Janssen." *Études photographiques,* no. 4 (1998).

Snyder, Joel. "Visualization and Visibility." In *Picturing Science, Producing Art,* edited by Caroline A. Jones and Peter Galison, 379–97. New York: Routledge, 1998.

Sokal, Alan, and Jean Bricmont. "Un Regard sur l'histoire des rapports entre science et philosophie: Bergson et ses successeurs." In *Impostures intellectuelles,* 165–84. Paris: Odile Jacob, 1997.

Staley, Richard. "Michelson and the Observatory." In *The Heavens on Earth: Observatory Techniques in Nineteenth-Century Science,* edited by David Aubin, Charlotte Bigg, and H. Otto Sibum. Durham, N.C.: Duke University Press, forthcoming.

———. "Traveling Light." In *Instruments, Travel and Science: Itineraries of Precision from the Seventeenth to the Twentieth Century,* edited by Marie-Noëlle Bourget, Christian Licoppe, and H. Otto Sibum. London: Routledge, 2002.

Stanley, Matthew. "An Expedition to Heal the Wounds of War: 1919 Eclipse and Eddington as Quaker Adventurer." *Isis* 94 (2003): 57–89.

Stengers, Isabelle. *The Invention of Modern Science.* Translated by Daniel
    M. Smith. Edited by Sandra Buckley, Michael Hardt and Brian Massumi.
    Theory Out of Bounds. Minneapolis: University of Minnesota Press,
    2000.

Sturgis, Alexander. *Telling Time.* London: National Gallery Company,
    2000.

Tacchini, P. "Sulla equazione personale." *Rivista Sicula di scienze, lettera-
    tura ed arti* 2 (1869): 382–92.

Taylor, Frederick Winslow. *Principles of Scientific Management.* New York:
    Harper and Brothers, 1911.

Terrall, Mary. "Representing the Earth's Shape: The Polemics Surrounding
    Maupertuis's Expedition to Lapland." *Isis* 83 (1992): 218–37.

Thomson, William. "Electrical Units of Measurement" (1883). In *Popular
    Lectures and Addresses.* London: Macmillan, 1891–94.

Tissandier, Gaston. "Les Éclairs reproduits par la photographie instan-
    tanée." *La Nature: Revue des sciences et de leurs applications aux arts et
    à l'industrie* (1884): 76–77.

Titchener, Edward Bradford. "Notes: Wundt's Address at Speyer, 1861."
    American Journal of Psychology 34 (1923): 311.

Touchet, Em. "La Forme et la structure de l'éclair." *La Nature: Revue des sci-
    ences et de leurs applications aux arts et à l'industrie* (1904): 138–40.

Trajectoires du rêve. Paris: Paris-Musées/Actes Sud, 2003.

Trouvelot, Étienne Léopold. "Cambridge, Mass.: Physical Observatory of L.
    Trouvelot, Esq." *Annual Record of Science and Industry for 1878* (1879):
    51–52.

———. "Étude sur la durée de l'éclair." *Comptes rendus des séances de
    l'Académie des sciences* 108 (1889): 1246–47.

———. "Étude sur la foudre." *La Lumière électrique* 29 (1888): 254–55.

———. "Étude sur la structure de l'éclair." *L'Astronomie* 7 (1888): 303–06.

———. "La Photographie appliquée à l'étude de l'étincelle électrique." *La
    Nature: Revue des sciences et de leurs applications aux arts et à l'industrie*
    (1889): 109–10.

———. "La Photographie appliquée à l'étude des décharges électriques."
    Comptes rendus des séances de l'Académie des sciences 107 (1888): 684–85.

———. "L'Éclipse de Soleil du 17 juin par M. E.-L. Trouvelot." *Comptes ren-
    dus des séances de l'Académie des sciences* (1890): 1322–23.

———. "L. Trouvelot's Physical Observatory, Cambridge, Mass." *Annual
    Record of Science and Industry for 1877* (1878): 41–42.

———. *Observations sur les planètes Vénus et Mercure.* Paris: Gauthier-
    Villars et fils, 1892.

————. "Sur la forme des décharges électriques sur les plaques photographiques." *La Lumière électrique* 30 (1888): 269–73.

————. *The Trouvelot Astronomical Drawings Manual.* New York: C. Scribner's Sons, 1882.

Trutat, Eugène. *Dix Leçons de photographie. Cours professé au Muséum de
Toulouse,* Bibliothèque photographique. Paris: Gauthier-Villars, 1899.

————. *La Photographie animée.* Bibliothèque photographique. Paris:
Gauthier-Villars, 1899.

Tucker, Jennifer. *Nature Exposed: Photography as Eyewitness in Victorian
Science.* Baltimore: Johns Hopkins University Press, 2005.

————. "Photography as Witness, Detective and Impostor: Visual Representation in Victorian Science." In *Victorian Science in Context,* edited by
Bernard Lightman, 378–408. Chicago: Chicago University Press, 1997.

Tucker, R. H., and Knight Dunlap. "The Personal Equation and Reaction
Times." *Science* 57, no. 1480 (1923): 557–59.

Ule, Otto Eduard Vincenz. "Sur les moyens de mesurer la pensée: Lettre de
M. Ule à M. E. Desor." *Revue suisse* 20 (1857): 197–202.

Van Helden, Albert. "Measuring Solar Parallax: The Venus Transits of
1761 and 1769 and Their Nineteenth Century Sequels." In *Planetary Astronomy from the Renaissance to the Rise of Astrophysics, Part B: The
Eighteenth and Nineteenth Centuries, the General History of Astronomy,*
ed. René Taton and Curtis Wilson. Cambridge: Cambridge University
Press, 1995.

Veblen, Thorstein. *Higher Learning in America* (1918). New Brunswick, N.J.:
Transaction, 1993.

Vitoux, Georges. *La Photographie du mouvement.* Paris: Chamuel, 1896.

Vlès, Fred. *La Cinématographie astronomique.* Bibliothèque générale de cinématographie. Paris: Charles Mendel, c. 1914.

Voltaire. "Micromégas: A Philosophical Story." In *Micromégas and Other
Short Stories,* 17–35. London: Penguin Books, 2002.

W. H. "Figures produites sur des plaques photographiques par des décharges
électriques, par J. Brown." *La Lumière électrique* 30, no. 51 (22 December
1888): 579–81.

————. "Figures produites sur des plaques photographiques sèches par les
décharges électriques, par J. Brown." *La Lumière électrique* 30, no. 42 (20
October 1888): 134–35.

Watson, John B. *Behaviorism.* New York: People's Institute Publishing Company, 1924.

Wise, M. Norton, ed. *The Values of Precision.* Princeton, N.J.: Princeton University Press, 1995.

Wolf, Charles. "Conférence sur les applications de la photographie à l'astronomie et en particulier à l'observation du passage de Vénus." *Bulletin de la Société française de photographie* 21 (8 January 1875): 16–28.

———. *Histoire de l'Observatoire de Paris de sa fondation à 1793.* Paris: Gauthier-Villars, 1902.

———. "La Figure de la Terre, soirées scientifiques de la Sorbonne." *Revue scientifique* 7 (12 March 1870): 226–34.

———. "Le Passage de Vénus sur le Soleil en 1874." *Revue scientifique* 9 (20 April 1872): 1006–11.

———. "Le Passage de Vénus sur le Soleil en 1874, conférence faite à la Société des amis des sciences, le 29 mai 1873." In *Recueil de mémoires, rapports et documents relatifs à l'observation du passage de Vénus sur le Soleil,* 377–401. Paris: Firmin Didot, 1874.

———. "Recherches sur l'équation personnelle dans les observations de passages, sa détermination absolue, ses lois et son origine." *Annales de l'Observatoire impérial de Paris* 8 (1866): 153–208.

———. "Sur la comparaison des résultats de l'observation astronomique directe avec ceux de l'inscription photographique." *Comptes rendus des séances de l'Académie des sciences* 102 (1 May 1886): 476–77.

Wolf, Charles, and Charles André. "Recherches sur les apparences singulières qui ont souvent accompagné l'observation des contacts de Mercure et de Vénus avec le bord du Soleil, Mémoire présenté à l'Académie des sciences, dans sa séance du 1er mars 1869." In *Recueil de mémoires, rapports et documents relatifs à l'observation du passage de Vénus sur le Soleil,* 115–72. Paris: Firmin Didot, 1874.

———. "Sur le passage de Mercure du 4 novembre 1868, et les conséquences à en déduire relativement à l'observation du prochain passage de Vénus." *Comptes rendus des séances de l'Académie des Sciences* 68 (25 January 1869): 181–83.

Wolf, Charles, and Antoine Yvon-Villarceau. "Rapport sur les mesures micrométriques directes à faire pour l'observation du passage de Vénus, présenté à la Commission du passage de Vénus, dans la séance du 8 mars 1873." In *Recueil de mémoires, rapports et documents relatifs à l'observation du passage de Vénus sur le Soleil,* 337–44. Paris: Firmin Didot, 1874.

Wolf, Stewart. *Brain, Mind, and Medicine: Charles Richet and the Origins of Physiological Psychology.* New Brunswick, N.J.: Transaction Publishers, 1993.

Woodbury, Charles H. *Painting and the Personal Equation.* Boston: Riverside Press, 1919.

Woolf, Harry, ed. *Quantification: A History of the Meaning of Measurement in the Natural and Social Sciences.* Indianapolis: Bobbs-Merrill, 1961.

Worms, Frédéric. "Entre critique et métaphysique: La Science chez Bergson et Brunschvicg." In *Les Philosophes et la science,* edited by Pierre Wagner. Paris: Gallimard, 2002.

Wozniak, Robert H. "Théodule Armand Ribot: German Psychology of To-Day." In *Classics in Psychology, 1855–1914: Historical Essays.* London: Thoemmes Continuum, 1999. Originally published 1879; first published in English 1886.

Wundt, Wilhelm. *Éléments de psychologie physiologique.* Translated by Élie Rouvier. Vol. 2. Paris: Germer Baillière et cie, 1886.

———. "Préface de l'auteur pour l'édition française." In *Éléments de psychologie physiologique,* xxvii–xxviii. Paris: Félix Alcan, 1886.

Zenger, Ch.-V. *Comptes rendus des séances de l'Académie des sciences* 109 (1889): 294–95.

INDEX